MW01351853

Royal Prussian Jagdstaffel 30

Its History

Its Pilots

Its Colors & Markings

Bruno Schmäling & Winfried Bock

Royal Prussian Jagdstaffel 30

Its History

Its Pilots

Its Colors & Markings

Bruno Schmäling & Winfried Bock

Interested in WWI aviation? Join The League of WWI Aviation Historians (**www.overthefront.com**) and Cross & Cockade International (**www.crossandcockade.com**)

Design and layout: Jack Herris
Cover design: Aaron Weaver
Front cover & interior painting: Russell Smith
Back cover painting: Jerry Boucher
Color profiles: Jim Miller
Digital photo editing: Aaron Weaver & Jack Herris
Printed by Walsworth, Marceline, MO, USA

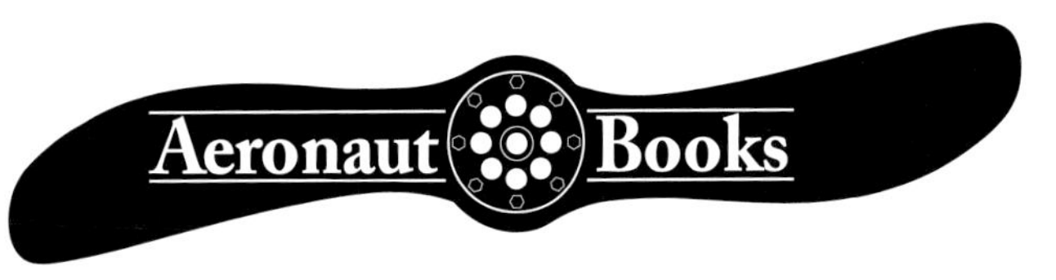

www.aeronautbooks.com

Publisher's Cataloging-in-Publication data

Schmäling, Bruno, and Bock, Winfried
Royal Prussian Jagdstaffel 30 / by Bruno Schmäling and Winfried Bock.
p. cm.
ISBN 978-1-935881-25-4
1. Fuchs, Otto, 1897 – 1987. 2. Jagdstaffel 30 3. World War, 1914–1918 -- Aerial operations, German. 4. Fighter pilots -- Germany. 5. Aeronautics, Military -- Germany -- History. II. Title.

ND237 .S6322 2011
759.13 --dc22 2011904920

Dedicated to Otto Fuchs
and the former members of Royal Prussian Jagdstaffel 30

Above: Otto Fuchs in his Albatros D V at Jagdstaffel 30 in the summer of 1917. His personal dedication reads: *To be means to be moved. Not to be stuck in space. What is time? It plunges into the abyss. Only that which we experience deeply within its womb gives zest to my present without change. Otto Fuchs to his dear Bruno J. Schmäling, the indefatigable reviver of days gone by.*

Right: Otto Fuchs as a student with his self-constructed model airplanes.

Contents

The Authors

Bruno Schmäling has dealt intensively with the history of the German air service in the First World War for many years and has already published numerous articles on this subject in Germany, England, and the USA.

Winfried Bock, Diplom-Ingenieur and Hauptmann d. R. of the Bundesluftwaffe (Captain of the reserve of the German Air Force), studied air and space technology and works for a European air and space firm. He is an author and co-author with over 20 publications about the history of German fighter aviation.

Contributors

A book such as this is never the result of the authors' efforts alone, but rather always represents the teamwork of a number of individuals. For this reason, our thanks are due to the generous and dedicated collaboration of the following:

Rainer Absmeier
The late Dr. Gustav Bock
The late Alex Imrie – pilot, aviation historian, mentor, and friend
Reinhard Kastner
Volker Koos
The late Les Rogers
Michael Schmeelke
Marton Szigeti
Hannes Täger
Aaron Weaver
Adam M. Wait
Greg VanWyngarden
Reinhard Zankl

Unless otherwise noted, all photographs are from the collection of Bruno Schmäling.

Acknowledgements

Our most sincere gratitude is due for the kind support to Prof. Dr. Dr. Holger Steinle and Mrs. Cathrin Clemens of the Stiftung deutsches Technikmuseum, Berlin, the Bayerisches Hauptstaatsarchiv, Abteilung IV, in Munich and the Stadtarchiv of the "Freien und Hansestadt Hamburg".

We would also like to convey our sincere thanks to Mrs. Christa von Schultzendorff (the daughter of Rudolf Freiherr von der Horst zu Hollwinkel), Mrs. Erika Imrie, Mr. Pieter Kaye (the son of Kurt Katzenstein), Mrs. Birgit and Mr. Jochen Kalin, Mrs. Maria and Mr. Hans-Joachim Erbguth, Prof. Dr. Frank Erbguth, Mr. Olaf von Bertrab, Mr. Georg Baron von Manteuffel-Szoege, Mr. Hans-Günther von der Marwitz, as well as Mr. Walter von Hueck (former head of the Deutsches Adelsarchiv) and Herrn Dr. Christoph Franke (head of the Deutschen Adelsarchiv).

Last, but very far from least, we would like to express our appreciation to Adam M. Wait for his translation of the German manuscript into English.

Bruno Schmäling & Winfried Bock

Forward

This is the first book in a series about German Jagdstaffeln of World War 1. While the history of the German air service in the First World War has almost been forgotten in Germany, it enjoys a great deal of interest especially in English-speaking countries. Accordingly, after the Second World War most published works concerning this topic were written above all by English and American aviation enthusiasts.

This planned series by German authors not only has the goal of supplementing these previous efforts with up to now unpublished original German documents, photos, and facts, but the attempt shall also be made to correct false clichés and phraseology, and in the light of the facts to straighten out one or the other popular legend which originated during the course of the past decades.

The primary foundation for this is the extracts from unit War Diaries prepared by Herr Erich Tornuss of Weissenfels in the 1930s. Despite all the gaps, these transcriptions are of inestimable value, as the original War Diaries – with two exceptions – were destroyed in the archive at Potsdam by Allied bombs during World War 2.

A further basis is provided by the still extant weekly reports of the Kommandeure der Flieger of the individual German Armies, aerial combat reports of former flyers, reports of the flak group commands, entries in aircraft spotting logs, documents from Armee-Flugparks and other official and semi-official documents of the German air service, as well as flight logbooks, letters, photos, and personal notes which were painstakingly collected and evaluated.

These documents were supplemented by information gained through conversations with former members of the air service conducted during the years 1958 to 1983 by Dr. Gustav Bock, Alex Imrie, and Bruno Schmäling, amongst others. Their statements, accounts, and anecdotes, as well as their characterizations of persons have also seeped into this work.

A further focus of the planned series is the markings and paint schemes of the aircraft. Due to a lack of information, there have previously been frequent attempts to reconstruct the paint schemes of German aircraft on the basis of so-called "grayscale interpretation" of the black-and-white photographs. Because there is no scientific foundation for this, the resulting specifications regarding the colors are purely speculative and often in error. The paint schemes described in this series, on the other hand, are mainly based on the statements of former fighter pilots or their mechanics.

These statements reveal that the manner in which aircraft were painted was closely connected with the personality of the pilot concerned, his social and regional origins, his military career, or his preferences, dreams, wishes, and hopes. Therefore there was often a purely personal reason or even a personal story behind the paint schemes. To the extent that these details are known, they will be presented in this series. This information is supplemented by official reports, for example reports by the flak group commands or personal records of contemporary witnesses.

Of course, in such a work as this there are almost inevitably gaps and open questions. For example, many photos show only a part of the aircraft, and so in the representation of the paint scheme the unseen portion must more or less be guessed at on the basis of available documents. Therefore the authors would be grateful for any sort of supplemental material or additional information.

Bruno Schmäling and Winfried Bock
Summer 2013

Above: Bruno Schmäling

Right: Winfried Bock

Prolog

Northern France in March 1917

The rust red and dark green painted airplane rumbled a few more meters over the grass-covered soil of the airfield near Houplin, about 8 km southwest of the city of Lille, and then it took off and slowly gained height in a long drawn out spiral. Some gusts of wind seized the two-seater Roland C IIa C.2701/16 and gave the two crew members a thorough shaking.

In the pilot's seat Otto Fuchs crouched down in order to protect himself from the ice-cold wind. Behind him his brother Rudolf had withdrawn deep into the observer's cockpit. The machine climbed in spirals up to about 2000 meters and then took a course towards the front.

Suddenly Rudolf tapped on his brother's shoulder from behind and pointed upwards to the right. An Albatros D III coming from above appeared almost playfully at their side. The rudder of the fighter plane was painted blue and edged in white. The Albatros briefly wagged its wings and already the single-seater fighter climbed to a position some hundreds of meters above the two-seater.

On the left side appeared a second Albatros D III, whose white rudder outlined in black and bearing the national markings shone in the sunlight. On the fuselage glared a large red "E." The pilot waved to the brothers and then likewise pulled his light single-seater into a position above the Roland.

The presence of the two fighter planes gave Otto Fuchs a feeling of security. The artillery-spotting aircraft of Flieger-Abteilung (A) 292 were often accompanied by fighter planes of the neighboring Jagdstaffel 30. The lightness and maneuverability of these aircraft impressed Otto and every time he encountered them his desire to also fly one of these slim and swift airplanes increased as well...

A few days later, on March 28, 1917, Otto Fuchs sat again in the pilot's seat of his "Walfisch" [whale] – as the Roland C II / C IIa was known – and made practice turns around the Abteilung's airfield. To the side of the airfield, white targets had been set up and Otto tried to hit them with the machine-gun installed in front of his pilot's seat. After he had finished the exercise, he briefly landed in order to take along one of the Abteilung's observers – Leutnant Wolf von Manteuffel-Szoege – on a sightseeing flight. This observer had his camera on hand in order to shoot a few photos in flight.

Otto Fuchs took off with the "Walfisch" and turned southeastwards, gaining height, in order to turn over Seclin and Lesquin before flying back to the airfield. To the west of Frétin he saw an apparently damaged Albatros losing height, levelling out, and coming in for a hard landing. A few moments later it crashed into a fence surrounding a garden and in the process wound up on its nose. Otto landed in the vicinity and had a car driving by take him and Wolf von Manteuffel-Szoege to the scene of the crash.

Upon his arrival Otto saw the pilot, who was wearing a greasy, oil-stained leather jacket and was apparently uninjured. He recognized the pilot as the commander of Jagdstaffel 30, Oblt. Hans Bethge. Otto learned from the excited people gathered around him that the Staffel commander had been attacked by several British Nieuports. After a difficult dogfight he had succeeded in shooting down the British pilot, but the latter had holed his fuel tank, forcing him to make an emergency landing. Otto Fuchs accompanied Hans Bethge to the nearby crash site of his opponent. Near the wreckage of the enemy Nieuport Otto wondered at Bethge's behavior:

For a while the Oberleutnant observes the pitiful remains. I conceal my surprise over the fact that he does not at all look like a victor. There is no sign of pride or the rush of joy which, so I think, must course through his veins like an intoxicating wine. In sober words he suggests that we go back. He wants to call his Staffel. I am stricken by a cold, mysterious look in his eyes.[1]

After his first aerial victory, as Otto Fuchs related many years later, he could well understand this behavior.

Hans Bethge offered to take Otto and Wolf to the spot where their Roland had landed. During the drive, Otto summoned his courage and told Hans Bethge of his desire to become a fighter pilot. When Hans Bethge found out that he flew his missions in a Roland C II, which was difficult to fly, he promised to lend him his support. A little later Otto and his brother visited Jagdstaffel 30 and afterwards Otto was a guest a few times at Phalempin. On May 29, 1917 he was ordered to the Jagdstaffelschule [fighter pilot school] located at Valenciennes. His training there lasted for only two days and then he was on his way to a Jagdstaffel, but in an entirely different manner than he had imagined.

Above: Otto and Rudolf Fuchs with their Roland C IIa C.2701/16 of Flieger-Abteilung (A) 292 in the spring of 1917 at Houplin airfield in northern France.

Right: The wreckage of Nieuport 17 A6615, flown by 2/Lt. H. Welch of 1 Squadron RFC, after the aerial combat on March 28, 1917. In the background the Albatros D III in which Oblt. Bethge made an emergency landing can be seen.

Below: The cockade shows where Nieuport A6615, flown by 2/Lt. H. Welch of 1 Squadron RFC, came down.

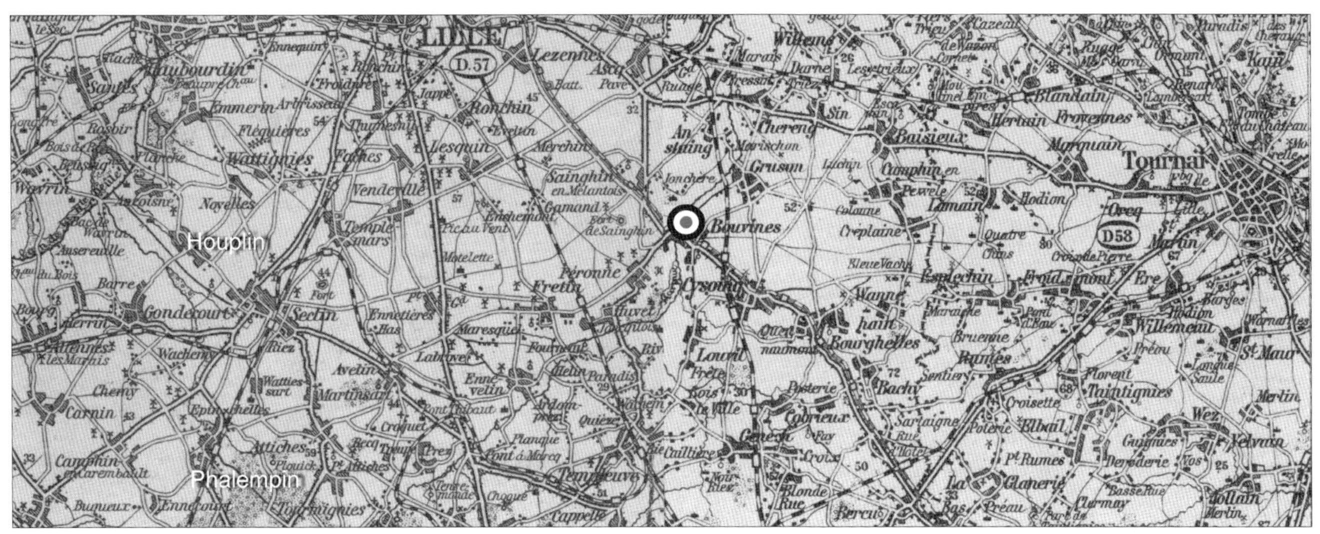

1. January 1917 – May 1917

Formation and First Weeks of Front-Line Duty

On December 14, 1916, Jagdstaffel 30 was established by order of the Kommandierender General der Luftstreitkräfte [commanding general of the air service], or the (Kogenluft). Flieger-Ersatz-Abteilung 11 [Flyer Replacement Unit 11], abbreviated (FEA 11), in Breslau was put in set up process for the formation. In the following weeks the entire ground personnel for the new unit were assembled and the entire equipment of the Staffel—with the exception of the aircraft—was provided for.

In accordance with Order No. 929/16 of the Feldflugchef (the air service commander's title before it was changed to Kogenluft.), the establishment of a Jagdstaffel consisted of 14 single-seater fighter aircraft, a Staffel commander, an "Offzier zur besonderen Verwendung" (lit.: Special Duty Officer, that means "Executive Officer for all non-flying duties") and 12 pilots, as well as the 114 non-commissioned officers and ranks of the ground personnel. In addition, the equipment also consisted, amongst other things, of three personnel transport automobiles, four trucks, a water heating truck, a lighting truck with generator, a mobile workshop, and two trailers.

Shortly before the completion of the set-up process Oberleutnant Hans Bethge, who was appointed as Staffelführer by the Kogenluft, arrived in Breslau in order to formally take charge of his Staffel.

Hans Bethge was born on December 6, 1890 in Berlin as the son of Kapitänleutnant Wilhelm Bethge. He spent his youth in Friedrichshafen on Lake Constance. Because he was refused entry into the navy due to his shortsightedness, he joined Railway Regiment 1 in Berlin as an officer cadet and received his commission in 1912.

At the beginning of the war he took to the field with this regiment. Amongst the tasks of his unit were the expansion and repair of the railway system, as well as tunnel construction and bridge demolition. After suffering an injury, he reported to the air service in 1915. In May of the same year he began training as a pilot at FEA 4 in Posen and in November 1915 he arrived at "Brieftauben-

Above: Oblt. Hans Bethge in his study at Phalempin in the summer of 1917.

Right: The railway station at Phalempin in the summer of 1917. (R. Zankl)

Abteilung-Ostende" ["Ostend Carrier Pigeon Detachment"]. This was the cover name for the Kampfgeschwader der Obersten Heeresleitung 1 [Bomber Wing of the Supreme Army Command 1], or Kagohl 1, which had been established for strategic bombing raids on enemy targets and was in the process of being organized.

As protection against enemy fighter planes, a few single-seat Fokker monoplanes were also allocated to Kagohl 1, and among the pilots assigned to fly them was Hans Bethge. On June 12, 1916 he was ordered to Kampfeinsitzer Kommando Bertincourt, in which several single-seat combat aircraft were assembled. In the course of the establishment of the first Jagdstaffeln, the Kommando was renamed Jagdstaffel 1 on August 22, 1916. After achieving three victories by December 1916 followed his appointment as commander of the newly established Jagdstaffel 30.

Otto Fuchs described Hans Bethge as he saw him upon arrival at the Staffel in June 1917 with the following words:

We meet Bethge in his room lying on the divan and reading. He is of small, stocky build. I notice his sharply protruding cheek bones and his broad, domed forehead. A couple of merry amber eyes reside in deep sockets and around his delicately formed mouth appear all sorts of amusing little wrinkles, first sarcastic, then painful, then cheerful in accordance with Bethge's lively mood. He could be 28 to 30 years of age. A bold clarity warms and illuminates his entire being. It forms his actual center.[1]

In the course of the month of January the first pilots arrived at the Staffel. These were: Leutnant Hans von Schell from the Jagdstaffelschule; Unteroffizier Hans Rody and Gefreiter Heinrich Schneider, previously with Jagdstaffel 21; Leutnant Gustav Nernst from Jagdstaffel 10; Unteroffizier Josef Heiligers from Armee-Flugpark 6 (Aflup 6); and lastly Leutnant Douglas Schnorr, who took over the duties of Offizier z.b.V. Finally, on January 21, 1917 the Staffel was able to report the successful completion of its formation.

Besides the Staffel commander, Lt. Gustav Nernst was the most experienced combat pilot of the Staffel. He was born on January 25, 1896 in Göttingen as the son of the later Nobel prize recipient in chemistry, Walther Nernst. After his training as a pilot he was transferred to Feld-Flieger-Abteilung 5 to fly single-seater combat aircraft and was immediately assigned to Kampfeinsitzer-Kommando III. On September 25, 1916 it was renamed Jagdstaffel 10, with which unit he achieved his first victory.

Vizefeldwebel Josef Heiligers was born on April 23, 1894 in Aachen. There are no documents available concerning his career before his transfer to Jagdstaffel 30. Otto Fuchs remembered him as a modest, courteous, and reserved man who maintained excellent control over his aircraft and was amongst the best and most reliable pilots of the Staffel.

According to Otto Fuchs, Lt. Douglas Schnorr was the ideal choice for the position of Special Duty Officer and accordingly enjoyed great popularity. He was considered to be a gifted organizer and provided the pilots with good food and good wine, as well as theater and music hall tickets. Besides that he spoke fluent English, as his mother was from England. His language abilities came in particularly handy when captured English flyers were guests with the

Left: Halberstadt D V D.420/16 after Gefr. Schneider's emergency landing in English hands. The national insignias on the fuselage were cut out by English souvenir hunters. (R. Zankl)

Staffel. He had lost a leg in an airplane crash and was therefore unable to fly with the Staffel on combat operations. Since he always complained that he no longer got to fly, Oblt. Bethge finally commandeered for him a discarded Halberstadt single-seater with which he could carry out flights in the rear area, which however ended several times with the airplane standing on its nose or flipping over.

On January 23, 1917 began the Staffel's three-day transport by rail to the Western Front, where on January 25 the airfield at Phalempin, located in the sector of the German 6th Army, was occupied. Phalempin lies approximately 12 km south of the city of Lille and when the war began had about 2,000 inhabitants. According to Otto Fuchs, it was a typical little northern French town in the coal mining region, comparable to small towns in the German Ruhr area. In his memory, Phalempin was not a particularly picturesque place, but for the members of Jagdstaffel 30 it soon became a second home.

On the other hand, the Forest of Phalempin lying to the east was beautiful. There one could go for extensive walks. The relations with the French civilians were correct and polite, but distant. Only Hans Bethge had somewhat closer contact with the locals, due to his good knowledge of the French language. For example, he befriended a mining engineer, so that the pilots were able to undertake some excursions into one of the mines.

The Staffel's airfield lay on the northern edge of Phalempin to the west of the railway line to Lille. The pilots lived in a stately mansion in the town which was surrounded by a beautiful park.

The sector of the German 6th Army at the beginning of 1917 stretched around 60 kilometers from Armentières in the north to the area around Arras in the south. In the north was the adjoining sector of the German 4th Army and in the south the sector of the German 2nd Army.

The German 6th Army sector at this time was rather quiet. The enemy soldiers sat opposite in well strengthened positions and for the time being there were no major military operations. Accordingly combat activity in the air was also rather slight. The task of the Staffeln primarily consisted of warding off approaching enemy airplanes, and attacking enemy aircraft directing artillery, as well as the protection of friendly two-seater aircraft of the Flieger-Abteilungen (Flg.Abt.). Besides Jagdstaffel 30, Jagdstaffeln 4 and 11 were deployed in the army sector in February 1917, operating out of the Douai airfield about 20 km south of Phalempin.

The mission of the Artillerie-Flieger-Abteilungen (Flg.Abt.(A)) was the ranging of one's own artillery fire unto the enemy positions. For this reason, these units' aircraft were the preferred target of enemy fighter planes. The newly formed Schutzstaffeln, which were likewise equipped with two-seater aircraft, were responsible for the protection of the artillery planes. The number of Schutzstaffeln was however still too few in the spring of 1917 to provide escort for all the missions flown by the Flieger-Abteilungen. Flieger-Abteilung (A) 292, located at the Houplin airfield, had no Schutzstaffel regularly assigned to it. For this reason, Jagdstaffel 30 at times took on this task.

Otto Fuchs, who at this time was a member of Flg.Abt. (A) 292, reported concerning the cooperation with Jagdstaffel 30:

When, due to the military situation or good weather, increased enemy aerial activity was to be reckoned with, our Abteilung commander telephoned the commander of Jagdstaffel 30 and informed him of our planned operations. When we took off, Jasta 30 was notified by phone, whereupon the fighter planes took off. These then provided escort for our planes.

Right: Oblt. Hans Bethge's damaged Albatros D III D.2051/16 after recovery at the end of March 1917 at Phalempin airfield. The upper wing, rudder, and nose damaged from flipping over can be clearly seen. (Stiftung Deutsches Technikmuseum Berlin/ Historisches Archiv)

Above & Right: Albatros D III 2038/16, which crashed over Phalempin airfield on March 17,1917. Gefreiter Rody lost his life in the process. (Erbguth family)

This escort meant that the fighter planes remained in the vicinity in order to provide protection for our two-seaters in the event of an attack. In this way, there soon existed a friendly relationship between Flg.Abt (A) 292 and Jagdstaffel 30, whose airfields were not far from one another.[2]

On February 8, 1917 the first aircraft were assigned to the Staffel from Armee-Flugpark 6. Due to a lack of newer types (the actually planned Albatros D III was still barred from frontline use due to technical problems), three Halberstadt D Vs were

Above: Oblt. Hans Bethge in front of his white-painted Albatros D III D.2147/16. The instructions for covering the wings are provided on the metal sign left unpainted. (Stiftung Deutsches Technikmuseum Berlin/ Historisches Archiv)

flown over to Phalempin with which the Staffel was to at first get in some flying.

The primary task of the pilots at first was to become acquainted with the battle zone as quickly as possible. For this the most important aid to orientation was the railway lines, with which the German and Allied pilots alike became familiar. Fighter pilots with whom one of the authors conducted interviews concurred that enemy airplanes as a rule had to follow the railway lines in order to orient themselves over enemy territory. If one flew one's own missions along the railway lines which led to the front, one could almost be certain of encountering enemy planes. Only when one had been employed in a sector of the front for a long time did one learn to orient oneself by other prominent points like church towers, streets, lakes, or woods.

Just how important the rail lines were as aids in orientation is shown by the official military maps issued to the German pilots. In these the railway lines are especially emphasized. In Jagdstaffel 30's operating area were located the following railway lines: Douai—Seclin—Lille, Lille—Wavrin—La Bassée, Lille—Pérenchies—Armentières, Wavrin—Armentières, Seclin—Gondecourt—Annoeullin, Wahagnies—Carvin—Annoeullin, and Ostricourt—Lens. Further aids to orientation were the Canal de la Haute Deule and the Canal de la Bassée and the roads Seclin - Lens – Arras and Lille – Armentieres – Bailleul.

On February 15, 1917 Gefr. Heinrich Schneider

Above: Lt. Paul Erbguth. (Erbguth family)

Above: Lt. Paul Erbguth in front of his Albatros D III D.2140/16. (Stiftung Deutsches Technikmuseum Berlin/Historisches Archiv)

took off in a Halberstadt D V for a practice and frontline orientation flight. As a result of the very cloudy, foggy, and hazy weather he lost his orientation and went over the lines. After being fired upon by enemy flak he had to make an emergency landing. British soldiers rushing over took him prisoner.

Ten days later, on February 25, Oblt. Bethge, Lt. von Schell, and Lt. Nernst proceeded to Aflup 6 to pick up the first longed-for Albatros D IIIs for the Staffel. While Hans Bethge and Gustav Nernst safely landed their two aircraft at Phalempin, Lt. von Schell flipped over while landing and had to be sent to the hospital; his airplane was likewise due for the repair shop.[3] So for the time being only two of the new fighter planes remained to the Staffel. These were taken by Oblt. Bethge and Lt. Nernst, the two most experienced fighter pilots of the Staffel. On March 1, 1917 it was presumably also these two who took off for the first operational flight of the Staffel.

Five days later there was a further strengthening of the unit with the arrival of Lt. Joachim von Bertrab and on March 10, 1917 Lt. Paul Erbguth was transferred to Jagdstaffel 30.

Joachim von Bertrab was the son of a forester and was born on February 18, 1894 in Sankt Andreasberg near Zellerfeld in the Harz region. On February 8, 1916 he was transferred to FEA 1 in Altenburg for pilot's training and in November 1916 arrived at Feld-Flg.Abt. 71, where he was able to acquire his first frontline experiences as a pilot. From there his path led via Aflup A to Jagdstaffel 30. Otto Fuchs described him as a go-getter in aerial combat, brave to the point of boldness, at times too spirited in his actions. Hans Bethge regularly admonished him to reign in his eagerness for combat.

Paul Erbguth was born on November 20, 1891 in Reichenbach in the Vogtland (Kingdom of Saxonia) as the son of a book dealer. He had acquired his first experiences as a pilot at the front in Kampfstaffel S 3, which at the beginning of 1917 was renamed Schutzstaffel 21. According to Otto Fuchs, Paul Erbguth enjoyed a great deal of popularity due to his kind and friendly nature. In addition, he was very interested and talented in technical matters. One regularly found him in a black mechanic's togs in the hangar, where he tinkered with his airplane alongside the mechanics. At 25 years of age he was older than most of the other pilots, for which reason Hans Bethge liked to consult with him in discussions. Through his conciliatory manner he often functioned as a mediator in differences of opinion amongst the pilots. In addition, he had a reputation as a circumspect and prudent flight commander, for which reason he led the Staffel in the air during Bethge's absence. He also owned a gramophone with a collection of records, which was set up in the mess and ran almost every day. In

Above: Lt. Gustav Nernst in front of his Albatros D III D.2124/16 in April 1917. (M. Szigeti)

Above: Lt. Gustav Nernst. (M. Szigeti)

return, his comrades regularly had to listen to his favorite song: "Die Liebe vom Zigeuner stammt (the love comes from the gipsy). "

On March 17th some of the Staffel's aircraft took off for a practice flight. While doing so, Uffz. Hans Rody, who on this day was flying the Albatros of Lt. Nernst, was rammed at a low height from behind by a Staffel comrade. His machine went instantly "on its nose" and crashed in a vertical dive. Rody died upon impact.

This accident makes it clear that flying in close formation (as a Kette or a Staffel) posed many difficulties for the pilots.[4] Apart from the novices (called "Häschen" [little hares] in the flyers' jargon of the day), even pilots with frontline experience also had their share of problems with it. This was quite simply due to the fact that for a long time the war in the air had been a war of "single combatants,"

Above: Unsuccessful landing of Albatros D III D.2126/16 at Phalempin airfield. (M. Szigeti)

Above: Five Albatros D IIIs of the Staffel in April 1917. From the left one can see Bethge's white Albatros D III 2147/16 and as the third aircraft Lt. von Bertrab's black D III. (Stiftung Deutsches Technikmuseum Berlin/Historisches Archiv)

regardless of whether one was flying a single-seater or two-seater. Only former members of the Kagohl were familiar with formation flying from their own experience and consequently had the fewest problems with it.

Eleven days after the fatal accident, on March 28, 1917, the Staffel was finally able to achieve its first victory. Hans Bethge had appeared on the airfield at 10:00 and waited for the low-hanging clouds to dissipate so that he could take off for the front. He reported regarding the subsequent events:

I'm sitting in the pilot's house and play a game of patience. I have my pilot's clothing brought and am about to slowly put it on when the report arrives: 'Six enemy flyers over La Bassée.' Right after that someone barges in: 'Herr Oberleutnant, they're coming over the field!' The motors clattered, the machines are all standing ready for take-off. I get into my crate and five Albatros fighters roar away. I didn't let the fellows out of my sight for an instant as we climbed higher and higher.

Finally we're at the same altitude and now approach the enemy. I was about to attack the last of them, then three turn and now began the wildest combat I have ever participated in in the air. Machines were dashing like mad every which way. Almost nothing but cockades, seven Nieuports and just two Albatros. Loop, sideslip, wild squiggling, attacks and fleeing and attack in a wild dash.

Well, I slip 300 meters because I nosed over, then I toiled upwards again, and then I see a Nieuport somewhat to the side. I quickly attacked him.

Above: Soldiers of the 44th Reserve Batallion surround Sopwith 1½-Strutter A1073 of 43 Squadron RFC, which landed on their sector of the front on April 5, 1917. (R. Absmeieri)

I moved in altogether without danger and at 50 meters both of my machine-guns rattled. At the very same moment he did a lightning quick turn and then... horribly the fellow's right wings flew off.[5]

Following the aerial combat, Hans Bethge had to make an emergency landing with hits in his Albatros, during which his airplane crashed into a fence and wound up on its nose. His opponent in the air fight, 2/Lt. Hugh Welch, died in the crash of his Nieuport.

The described combat occurred, like most

Above: The officers of Jagdstaffel 30 on April 19, 1917 in front of the entrance doors of their quarters in Phalempin. In front, from left: Lt. von Schell, Lt. von Bertrab, Lt. von der Marwitz. Behind, from left: Lt. Brügman and Oblt. Bethge. In the rearmost row, from left: Lt. Erbguth, Lt. Nernst, Lt. Seitz, and Lt. Schnorr. (Erbguth family)

German victories over enemy aircraft, above German territory. This was for the following reason: The British pursued an "offensive strategy" in the conduct of the air war, which aimed at the British Royal Flying Corps fighting the German air service in its own airspace. This also proceeds from an British order captured by German troops, in which it says:

Pilots must continuously consider first of all that it is their duty to always fight above the enemy lines in an easterly direction and do their utmost to prevent the enemy from approaching even his own lines.[6]

This offensive strategy and the numerical superiority of the Allied air services led to the result that in most cases the German aircraft would become involved in aerial combats over their own territory, even before they had reached the front. Therefore the German flying units in the British sector were put under great pressure, but on the other hand this very strategy led to considerable losses in the British air service. While damaged

Right: The crew quarters building of the pilots of Jagdstaffel 30 in Phalempin. The structure on the roof of the building served the Staffel as an observation platform for the aircraft spotting service.

Right: The quarters of the members of the Jagdstaffel 30 in Phalempine, view from the street.

German aircraft could make an emergency landing in their own territory and the crews, insofar as they were not wounded, could be employed again the next day, an emergency landing for British pilots automatically meant becoming a prisoner of war. In addition, German aircraft which had been attacked or damaged could evade the attacks of Allied planes over their own territory by diving, since the latter usually gave up the pursuit at the latest when they had reached 1500 meters altitude, in order not to come within range of German anti-aircraft fire.[7]

Two days after the Staffel's first aerial victory, Lt. Heinrich Brügman arrived at the Staffel from Flg. Abt. (A) 240. He had been born in Mönchengladbach and grew up in Trier. Otto Fuchs characterized him as a friendly and delicate young man always in good spirits and in the mood for jokes, who regarded aerial combat more as "playing Indians" than as war. His nickname in the Staffel, alluding to his original regiment, was "Dragonerchen" (the little dragoon).

The Easter Offensive at Arras

The British army command planned a large scale offensive against the German positions on both

Above: Maintenance work is performed on the Albatros D III D.2305/16 of Lt. Hans von Schell in April/May 1917. (Stiftung Deutsches Technikmuseum Berlin/Historisches Archiv)

sides of the city of Arras in the spring of 1917. In contrast to the Somme offensive of 1916, the goal of the British attack was not just a breakthrough in the German lines. Rather, the British offensive was to tie down German forces on this part of the front so that a few days later in the southeast the French army on the Aisne could break through the German lines to the north of Reims. In this way the entire German front between Arras and Reims was to be cut off in a pincer movement. The area of attack for the British fighting forces in this operation stretched from Loos in the north to Fontaine in the south and had a width of about 35 km.

As preparation for their attack, on March 20th, 1917 the British army began with a constantly increasing artillery bombardment of the German positions. This intensified to an artillery duel, until on April 3, 1917 the British drumfire began.

On April 9, 1917 the British soldiers climbed out of their trenches and began the assault on the German positions. The British attackers succeeded in penetrating the German positions to about five or six kilometers, however the breakthrough of the German lines did not succeed.

From April 20 to April 22, 1917 followed a second period of drumfire as preparation for the next attack on April 23, 1917, which led however to only slight gains in territory in the area of Gavrelle—Roeux. These gains in territory were for the most part lost again during the German counterattack on the following day. The third and fourth attacks on April 28 and May 3, 1917 failed entirely.

After the attack preparations of the British had become discernible, the German fighter forces in the region of the German 6th Army were increasingly strengthened to ward off the British flyers and on the

Above & Right: Two in-flight photos of Albatros D III D.2126/16. In the photo at left the personal marking "X" has been painted on it. The marking indicates that the machine was flown by Vzfw. Heiligers. (G. VanWyngarden)

evening preceding the battle showed the following units:
Jagdstaffel 3 in Guesnain
Jagdstaffeln 4 and 11 at Douai
Jagdstaffel 27 at Tourmignies (as of April 19 at Bersée)
Jagdstaffel 28 at Wasquehal (transferred with the Gruppe Lille to the 4th Army on May 1)
Jagdstaffel 30 at Phalempin

During the course of the battle, further reinforcements were added:

Jagdstaffel 33 to Villers au Tertre on April 11
Jagdstaffel 12 to Epinoy on April 12
Jagdstaffel 10 to Bersée on April 28

As needed, the described forces were further supported by the adjoining German 2nd Army to

 The Airfields of the German Jagdstaffeln of the 6th Army in April 1917

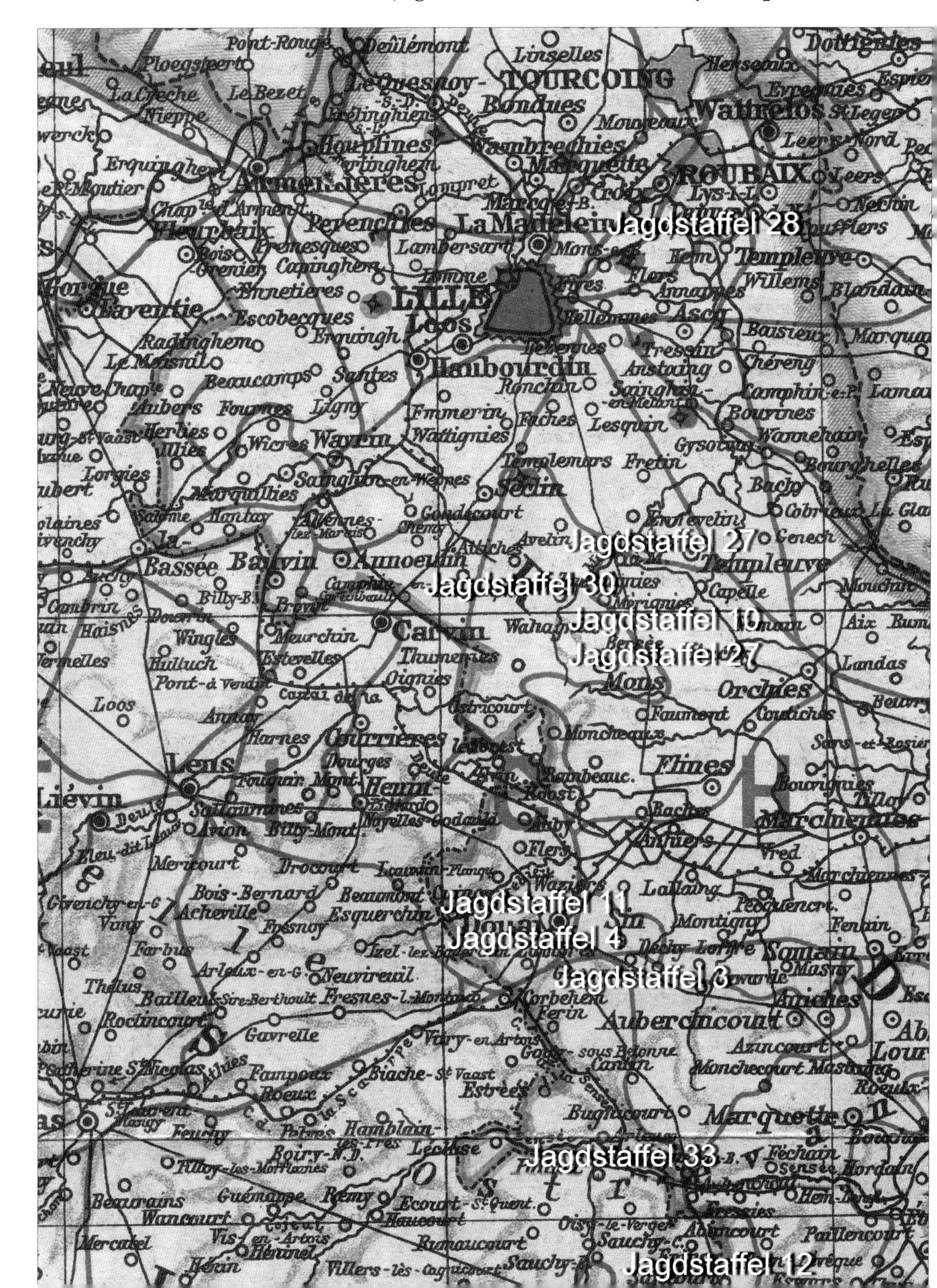

Above: The black men of the ground crew work on Albatros D III D.2306/16. (Stiftung Deutsches Technikmuseum Berlin/Historisches Archiv)

the south, which in April 1917 had at its disposal Jagdstaffeln 2, 5, 6, 20, and 26.

The British air service, which had a numerical superiority of about 5:1, attempted with a massive deployment of aircraft to keep their German opponents away from the zone of attack. But in contrast to the British offensive on the Somme in 1916, the Royal Flying Corps met with a well-organized fighter defense, which in addition had at its disposal a superior fighter plane in the form of the Albatros D III. In the course of bitterly contested air battles the German fighter pilots succeeded very quickly in gaining local aerial superiority and inflicting high losses on the British.

From the beginning of April to May the Staffeln listed reported the downing of a total of 260 enemy airplanes. In April 1917 alone the British flying units lost 160 machines over German territory and in the first half of May a further 48 airplanes. It is therefore no wonder that this month went down in the annals of British aerial warfare as "Bloody April."

On April 3, 1917 several aircraft of Jagdstaffel 30 took off amidst low-hanging clouds and stormy winds and flew at a low altitude in the direction of Douai, from where enemy aerial activity had been reported. South of Douai a flight from the Staffel spotted some Nieuport single-seaters of 40 Squadron, RFC at 14:40 and attacked them. Lt. Gustav Nernst was able to place himself behind an British machine and open fire. The enemy aircraft went down trailing a streamer of smoke behind it and eventually landed near Esquerchin, whereupon it went up in flames.

The pilot, 2/Lt. S. A. Sharpe, was able to jump out of the airplane uninjured and reach safety. German soldiers took him captive. Two days later Lt. Nernst was again successful. He attacked a Sopwith two-seater of 43 Squadron, RFC. During the exchange of fire the observer in the Sopwith was fatally wounded and the pilot succeeded in landing the machine near Rouvroy. He too was taken prisoner by German soldiers.

The next day was to be up to then the most successful day of the Staffel. Because there were very few clouds, already in the morning there reigned extremely lively enemy aerial activity. Around 7:00 a Kette took off due to the report of enemy airplanes crossing the lines and in the area behind the front in the vicinity of Bouvignies, about 10 km northeast of Douai, ran across a formation of Martinsyde bombers, which beforehand had bombed the small town of Ath. Already in the first attack Lt. von Bertrab was able to achieve a victory. While

Above: Manfred von Richthofen visiting Hans Bethge in April 1917. The reason for the visit was the coordination of the deployment of Jagdstaffeln 11 and 30 during the Easter offensive near Arras.

following the formation towards the direction of the front, he succeeded in gaining another victory just half an hour later. After that the Germans had to turn back because they were running low on gas.

At 10:20 the anti-aircraft battery located near Sainghin reported eight Sopwiths over La Bassée at 3500 meters altitude, whereupon the Staffel took off. Around 10:30 the British formation, which consisted

Right: Appraisal of Lt. Oskar Seitz, Bavarian FEA 1, 20 October 1916 (Bayerisches Hauptstaatsarchiv, München, OP 12233)

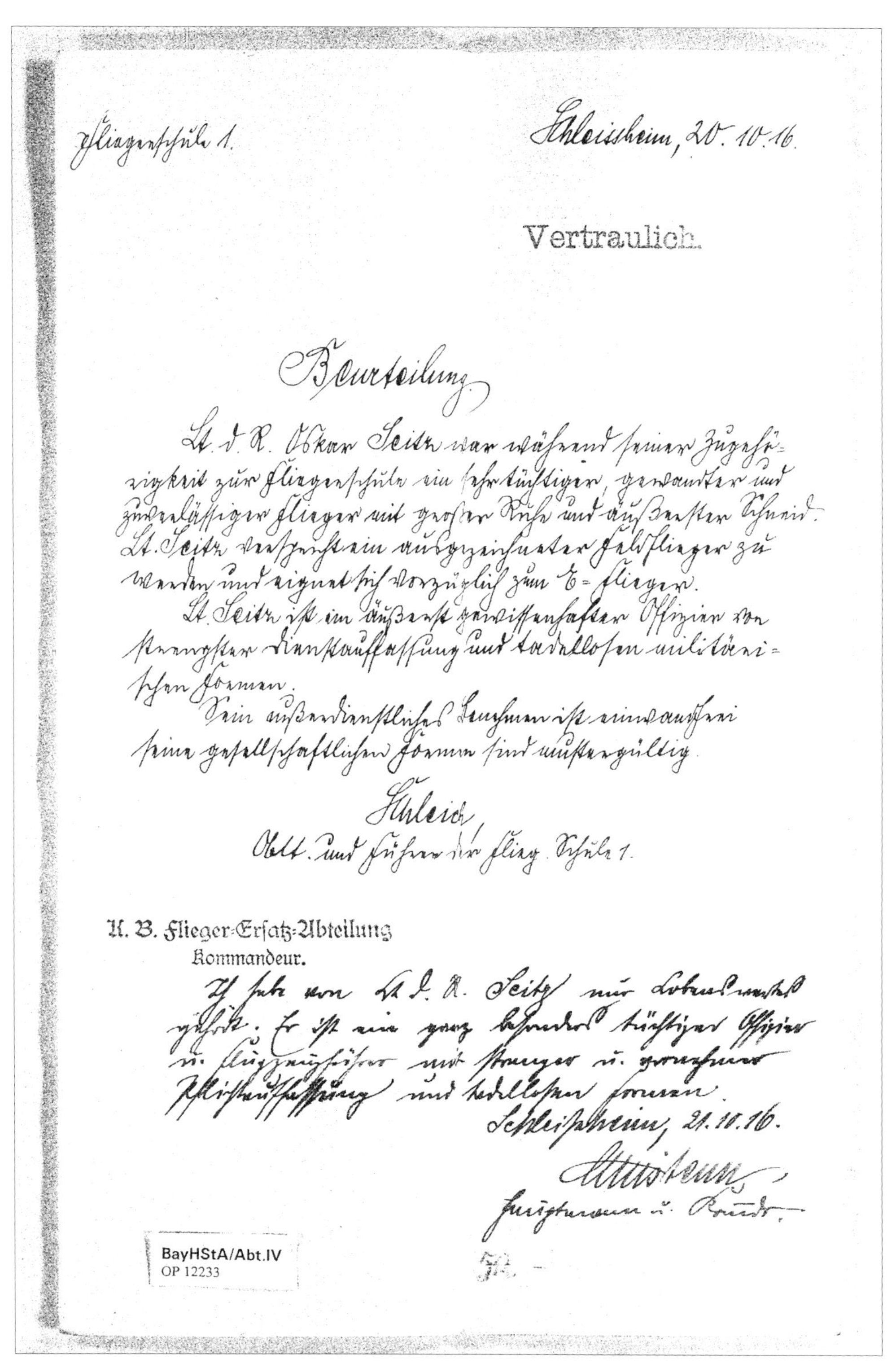

Fliegerschule 1. Schleissheim, 20. 10. 16.

Vertraulich.

Beurteilung

Lt. d. R. Oskar Seitz war während seiner Zugehörigkeit zur Fliegerschule ein sehr tüchtiger, gewandter und zuverlässiger Flieger mit großer Ruhe und äußerster Schneid. Lt. Seitz verspricht ein ausgezeichneter Feldflieger zu werden und eignet sich vorzüglich zum E-Flieger.

Lt. Seitz ist ein äußerst gewissenhafter Offizier von strengster Dienstauffassung und tadellosen militärischen Formen.

Sein außerdienstliches Benehmen ist einwandfrei, seine gesellschaftlichen Formen sind mustergültig.

Schleich,
Oblt. und Führer der Flieg. Schule 1.

K. B. Flieger-Ersatz-Abteilung
Kommandeur.

Ich habe von Lt. d. R. Seitz nur Lobenswertes gehört. Er ist ein ganz besonders tüchtiger Offizier u. Flugzeugführer mit strenger u. gewissenhafter Pflichtauffassung und tadellosen Formen.

Schleissheim, 21. 10. 16.

[illegible]
Hauptmann u. Kommandeur.

BayHStA/Abt.IV
OP 12233

of Sopwith 1½-Strutters of 45 Squadron, RFC, found itself above Tournai flying in the direction of Lille and Roubaix.

The Germans spotted the formation between Roubaix and Tournai and attacked. Here Sopwiths A1093 and A2381 came under the fire of Lt. Joachim von Bertrab's machine-guns, one shortly after the other. Both British aircraft crashed in short succession to the north and south of Pecq. Hans Bethge had attacked Sopwith 7806 and pursued the British pilot up to the railway station at Nechin near Tournai and shot him down in flames. The rest of the British formation was able to escape over their own lines. Upon their return the British pilots reported the downing of two German airplanes. Since no German losses can be determined, perhaps

Above: Five Albatros D IIIs of the Staffel at Phalempin airfield in May 1917. Recognizable are the Albatros D III D. D.2304/16 of Lt. Seitz (far right), next on the left the Albatros D.2305/16 of Lt. von Schell, and behind that the wine red D III of Lt. von der Marwitz. The old Staffel or "Ketten" insignia (rudder with black edge) is still to be seen on two Albatros D IIIs. The Albatros D V discernible at far left with the black rear fuselage is the aircraft of Oblt. von Tutschek, the Staffelführer of Jagdstaffel 12. The small table standing in the foreground served as a place for the field telephone while on standby duty, through which the order to take off was conveyed. (Stiftung Deutsches Technikmuseum Berlin/Historisches Archiv)

in the heat of the fight crashing Sopwiths were confused with Albatros.

The Staffel had reason to celebrate: five incontestable victories without a single loss. In addition, with his four victories in one day Lt. von Bertrab had achieved an accomplishment which no other German flyer before him had attained. It remains to be added that Lt. Joachim von Bertrab would have to share his record with Lt. Wilhelm Frankl of Jagdstaffel 4; for he too succeeded in downing four enemy aircraft on the same day, one of them at two o'clock in the morning as the first night fighter victory in the history of aerial warfare.[8]

On April 16, 1917 Leutnant von Schell spotted a Sopwith in combat with a German two-seater near Douai. He came to the aid of the German airplane and together they were able to force the British pilot to land on the German side near Douai. However, the victory was credited to the crew of the two-seater, who were from Schutzstaffel 4.

On the same day Lt. Oskar Seitz arrived at the Staffel. He was born in Mannheim in Baden on January 22, 1892. On October 1, 1913 he joined the 5. Bayerisches Chevaulegers Regiment "Erzherzog Friedrich von Österreich" in Saargemünd in Lorraine. On March 12, 1915 he was promoted to Leutnant der Reserve. In the same year he was awarded the Bavarian Military Order of Merit 4th Class with Swords. In the spring of 1916 he joined the air service and was sent for his training to the Bavarian FEA 1 at Oberschleissheim.

The commander of Fliegerschule 1 provided the following evaluation of Lt. Oskar Seitz after the completion of his training:

While attending the flying school Lt. Oskar Seitz was a very capable, skillful, and reliable flyer who showed great calm and the utmost nerve.

Lt. Seitz shows promise of becoming an excellent military aviator and is superbly suited for flying E-type aircraft [single seat monoplane fighters—authors]. Lt. Seitz is an extremely conscientious officer with the highest concept of duty and flawless military deportment. His off-duty conduct is irreproachable. His social manners are exemplary. (signed) Schleich, Oblt. and commander of Flieger-Schule 1.[9]

On November 1, 1916 he was transferred as a pilot to Bavarian Feld-Flieger-Abteilung 5, which in the

Above: Pilots of the Staffel with a ration of fresh bread. From left: Lt. Erbguth, Lt. von der Marwitz, Lt. von Schell, Lt. Schnorr, Lt. von Bertrab (Stiftung Deutsches Technikmuseum Berlin/Historisches Archiv)

same month was renamed Flieger-Abteilung (A) 292.

After successful missions with this unit, he reported for duty as a fighter pilot in the spring of 1917. He was transferred to Jagdstaffel 30 and then assigned to the „Kette" led by Lt. Nernst. On the following day he received the pilot's badge.

Two days later a pilot arrived at the Staffel who was to have a decisive influence on the history of the Staffel up to the end of the war. Lt. Hans-Georg von der Marwitz came from Ohlau in Silesia, where he was born on August 7, 1893. His great-grandfather was Friedrich August Ludwig von der Marwitz (1787–1837), a famous general from the time of the Wars of Liberation. In accordance with his family's tradition, he too entered into a military career and joined Ulanen Regiment Hennings von Treffenfeld (Altmärkisches) Nr. 16. As there were fewer and fewer tasks for the cavalry, he reported for training as a pilot and in March 1916 arrived at FEA 4 in Posen.

He experienced his first missions as a pilot with Kampfstaffel 28 in Kampfgeschwader 5 (Kagohl 5), which after the dissolution of the Kagohls was renamed Schutzstaffel 10. There he achieved his first victory, over a Farman near Bouchavesnes, on January 5, 1917. Otto Fuchs described the elegant, dashing cavalryman as a nobleman in the truest sense of the word: full of energy, brave, absolutely reliable in aerial combat, and endowed with a suitable portion of humor. He was a born leader, possessed charisma, and next to Hans Bethge was the best pilot and marksmen in the Staffel.

He was also known for taking on novices in the Staffel and protecting them during their first operational patrols. Like Hans Bethge, while on missions Hans-Georg von der Marwitz displayed courage and determination combined with circumspection and a careful weighing of risk. According to Fuchs, off-duty he was a man who seldom worried and took things easy. He had charm,

Left: German soldiers prepare the F.E.2d A6446 *"Malaya 17 The Alma Baker No. 3"* for transport. The F.E.2d was a so-called "presentation aircraft." These were airplanes which were financed through donations by private citizens or companies, whereby the donors had the right to give the airplane a name of their own choosing. (R. Absmeier)

quickly fell in love with this pretty woman or that, and enjoyed the nice aspects of life.

On April 20, 1917 Oblt. Hans Bethge began a four-week period of home leave and turned over the acting command of the Staffel to Lt. Paul Erbguth.

On the morning of April 21, 1917 low-hanging clouds and fog restricted to a large extent the deployment of aircraft in the morning and early afternoon. When the weather improved somewhat, two Ketten of the Staffel took off around 18:00 for the front, where lively enemy aerial activity had been reported. Lt. Gustav Nernst, who led one of the Ketten, was flying the white Albatros D III of Oblt. Bethge, since his own machine was not ready for operations.

Near Arras the Staffel encountered a British formation of F.E. 2bs from 25 Squadron RFC and Sopwith Triplanes of 8 Squadron RNAS. Lt. Nernst and Lt. Seitz apparently tried to attack the same enemy machine in the developing dogfight, whereby Lt. Nernst was rammed by the Albatros of Lt. Oskar Seitz passing over his machine at 18:45. With a badly damaged upper wing, his machine afterwards spun down on the other side of the lines near Arras. Gustav Nernst was killed upon impact. Oskar Seitz succeeded in making an emergency landing in a glide with his damaged machine in no man's land near Gavrelle, during which the airplane was wrecked. Under the protection of nightfall he was able to struggle back to his own lines on foot. He came away from this experience with a trauma later described as "a shock to the nerves".

In the war diary of the Royal Flying Corps it is noted regarding this aerial combat:

"2/Lt. R G Malcolm and 2/Lt. J B Weir of 25 Squadron attacked a formation of five enemy fighter planes. While doing so the F.E.was supported by F/Cdr. A.R. Arnold of 8 (Naval) Squadron. One of the enemy machines fell apart after it was badly hit and crashed within our lines.[10]

The British crews further reported the downing in flames of a second Albatros, as well as a third downed aircraft whose impact upon the ground was observed. The last-mentioned Albatros might have concerned the crashing machine of Lt. Seitz. There is no information regarding a German airplane shot down in flames at this point in time.

In Gustav Nernst the Staffel lost the most experienced "Kettenführer" besides Hans Bethge, who had in addition enjoyed great popularity within his circle of comrades. After his brother Rudolf had already fallen in 1914, the family had lost both sons in the war. Otto Fuchs reported that Oskar Seitz never got over this misfortune. Although Hans Bethge declared him free of any sort of blame and afterwards—as proof of his confidence in his flying abilities—put him in his Kette, Oskar Seitz was less and less able to cope with the demands of pursuit flying.

Three days later, on April 24, 1917, with good

weather and good visibility there came heavy aerial fighting. British fighter planes tried to prevent the invasion of the airspace over the battlefield by German airplanes up to an altitude of 5500 meters. At the same time reconnaissance and bomber aircraft penetrated German airspace in the rear area. At 07:45 Lt. Joachim von Bertrab and Uffz. Josef Heiligers each attacked a British B.E. over German-held territory between Douai and Valenciennes and according to their testimony shot them down. However, both victories were not confirmed.

On April 26, 1917, despite low-hanging clouds and intermittent rain there was lively enemy aerial activity. With his flight, Lt. Paul Erbguth attacked a B.E. of 10 Squadron RFC over German territory to the southeast of Haisnes and forced it to land. The wounded British crew, 2/Lt. F Roux (pilot) and 2/Lt. H J Prince (observer), were captured by German troops.

On May 1, 1917 Lt. Paul Erbguth and Lt. Heinrich Brügman each reported the downing of a Sopwith over German territory between Douai and Hénin-Liétard. Once more the victories were not confirmed. A week later it was Vzfw. Franz Bucher, newly transferred to the Staffel, who reported a victory over a Sopwith in the British lines near La Bassée. Lacking witnesses, this victory too was denied confirmation.

On May 13, 1917 Lt. Hans-Georg von der Marwitz and Lt. Paul Erbguth took off at 14:00 for an operational patrol upon the initiative of Hans-Georg von der Marwitz. The flight had a special background: in the spring of 1917 Hans-Georg had a girlfriend in Vienna with whom he was very much in love. He used every period of leave to visit her. Already shortly after his arrival at the Staffel in April 1917 he had submitted a request for leave to Hans Bethge, who flatly refused it with the indication that other pilots had a much greater right for leave than someone who had just arrived.

When on the day of Hans Bethge's departure he once more requested leave, the former remarked with annoyance that he would have to shoot down an enemy captive balloon before receiving permission for leave. It is not certain whether this offer by Hans Bethge was meant seriously, but in keeping with his nature, Hans-Georg took his Staffel commander's statement at face value and was immediately all afire at the thought of simply shooting down such a "gas bubble" quickly and then visiting his girlfriend.

He immediately gathered information regarding the locations of the captive balloons and "by chance" carried out his operational flights in the vicinity of the English observation balloons. Finally he discussed his plan with Paul Erbguth. Both were good friends, partly because Paul again and again functioned as a lender to Hans-Georg, who was regularly short on funds. Besides that, in Bethge's absence he was also entrusted with the leadership of the Staffel and therefore had to give formal approval for the balloon attack. On this occasion, Hans-Georg could also ask his friend right away whether he would accompany him on this adventure. Paul Erbguth was immediately prepared to do so. Now they just had to wait for favorable weather.

Above: Lt. Hans von Schell.

On May 13, 1917 the weather conditions were favorable for a balloon attack. Up till noon the sky became more and more cloudy and it rained. Hans-Georg von der Marwitz and Paul Erbguth took off together in the direction of Lille and near Bailleul

Above: "Sanke card" of Lt. Joachim von Bertrab.

crossed the English front with the protection of the cloud layer. Using the heavy clouds as cover, they approached the Allied captive balloons hanging between Dixmude and Ypres from the west unnoticed. Both dove together out of the clouds upon a captive balloon and fired upon it at close range.

Immediately little red flames appeared on the balloon. The two observers immediately bailed out and floated to the ground with their parachutes. While Paul pulled his Albatros upwards in a short spiral in order to keep a look-out for possible enemy planes, Lt. von der Marwitz attacked the balloon again, which thereafter went up in a fireball.

Now began for both of them the most dangerous part of the flight. Everything which could shoot on the Allied side went to work. Surrounded by clouds of flak, flying in a zigzag, the two finally succeeded in crossing over the front undamaged. Having arrived back at Phalempin, the downing of the captive balloon had already been reported, and both pilots were greeted with a big hullabaloo. Paul Erbguth renounced any claim for the victory with the words: "Now you can visit your girlfriend." The aerial victory was therewith granted to Lt. von der Marwitz.

When Hans Bethge returned from his leave a week later, he in fact declared Hans-Georg von der Marwitz to be "completely crazy," but kept his word and granted the leave. A beaming Hans-Georg was able to travel to see his girlfriend in Vienna.[11]

It must be explained in this connection that an attack on a captive balloon counted amongst the airmen's most risky endeavors in World War 1. Artillery fire was directed not only by aircraft, but also by captive balloons. If a captive balloon were shot down, it would take at least one to two, if not three, days until a new balloon had been prepared for deployment and ascended at that spot ready to take over the task anew. At this time the artillery was practically "blind" and therewith largely ineffective. For this reason, the captive balloons so extremely important for directing fire on both sides were effectively protected by at least by numerous flak batteries and machine-gun positions on the ground. In the French air service even fighter squadrons were sporadically employed for this task.

Two days later two Ketten of Jagdstaffel 30 took off after an incursion by two F.E.2ds were reported. After both F.E.2ds—they belonged to 20 Squadron—had dropped two 20 pound bombs without success on the winch of the captive balloon stationed near Quesnoy, they were attacked by the Albatros of the Staffel at 08:30. 2/Lt. Grout (F.E.2d A6446) was the first to recognize the German fighter planes and fired off a red signal cartridge. Thereafter the two F.E.s turned in a westwardly direction in order to escape over their lines.

According to Lt. Solly (F.E.2d A6354), they were each attacked by two German fighter planes, while two others remained above them and sought to cut off their path back over the lines. Hard-pressed by the six Albatros, the two F.E.s were forced in the direction of the German rear area and separated from one another. Lt. Solly last saw the aircraft of 2/Lt. Grout, pursued by two fighters, finally go down out of control.

Lt. Solly's machine-gunner later reported that the British machine went straight down and landed intact. In this connection he also claimed to have seen the crash of a German machine, but Jagdstaffel 30 did not have any losses. While the crew of F.E.2d

A6446 was taken prisoner by German soldiers unwounded after the landing, the second F.E. was able to escape over the British lines despite pursuit by an Albatros. The victory was credited to Lt. von Bertrab.

On the same day another pilot arrived in the person of Vzfw. Hans Oberländer to strengthen the Staffel. He was born on September 3, 1894 in Helmbrechts in Upper Franconia/Bavaria. From July to December 1916 he was on operational duties with Feld-Flieger-Abteilung 71. Together with his observer, name unknown, he caused a Nieuport to crash south of Azannes on October 25, 1916. Otto Fuchs characterized him as one of the best pilots of the Staffel, a man full of drive, but prudent and very reliable.

Oblt. Hans Bethge returned from his leave on May 20, 1917 and again took over leadership of the Staffel. On the same day Gefr. Josef Funk, from Zwiefalten in Württemberg, was transferred from Aflup.6 to the Staffel. The day after, besides new Albatros D IIIs, the Staffel was assigned the first four of the new D Vs from the Armeeflugpark. While ferrying over one of the new machines, Lt. Hans von Schell crashed while landing at Phalempin airfield once again and had to be sent to the hospital with severe injuries. His service with Jagdstaffel 30 was therewith ended. He did in fact return to a Staffel after his recovery, but was no longer fit for frontline service. A little later he went to Jagdstaffel 37 as Special Duty Officer.

On May 26, 1917 despite good weather and a cloudless sky there was only very little aerial activity in the morning. In the afternoon enemy aerial activity became more lively, but only during the last flight of the day was Oblt. Bethge successful. At 20:55 he forced a Nieuport of 29 Squadron, RFC to land on German-held territory to the east of Esquerchin. The pilot, 2/Lt. G M Robertson fell into captivity wounded.

For the month's finale, Gefr. Josef Funk shot down a captive balloon in flames near Nieppe at 10:10 on May 31, 1917.

Regarding Jagdstaffel 30's service, the weekly report of the Kommandeur der Flieger of the 6th Army recorded on June 1, 1917:

"From May 25 to 31 the Staffel undertook a total of 113 operational flights with a total duration of 105 flight hours, in the course of which there were 34 aerial combats."[12]

The report shows that only in about a third of operational flights was there aerial combats. This was a normal ratio for the circumstances in this sector of the front after the Easter battle. Many frontline patrols took place entirely without contact with the enemy or the enemy planes sighted succeeded in avoiding an aerial combat.

Markings and Paint Schemes of the Aircraft

Due to a lack of available conclusive photographic material, a Staffel marking or paint scheme for the Halberstadt D Vs initially employed by the Staffel cannot be determined and is not very probable, as these were a transitional solution until the arrival of the actual equipment. The first personal markings as well as a sort of Staffel marking were present on the Albatros D IIIs assigned as of the end of February 1917.

The Albatros D III D.2051/16 of Oblt. Bethge had a blue rudder with a white edge (Profile 1). On March 27, 1917 he took over D.2147/16, whose fuselage, upper surface of the top wing, and elevators he had painted white in order to be easily recognizable as the Staffel commander in the air (Profile 2). When Hans Bethge returned from leave in the middle of May 1917, he took over a new Albatros D III, which again received the blue-and-white rudder as a personal marking.

The Albatros D III D.2038/16 taken over by Lt. Gustav Nernst had a red "N" in front of the national insignia on the fuselage, the rudder had a black border, the vertical fin was a medium green with white border and the horizontal stabilizer was likewise painted green (Profile 3). Albatros D III D.2124/16, flown by Lt. Nernst in April, was painted in medium green on the upper surface of the upper wing and on the propeller hub.[13] The tail was likewise medium green and had a white edge. The green-and-white paint scheme was a reference to his original unit, the Königlich Preußisches Kürassier Regiment Graf Gessler (Rheinisches) Nr. 8 (Royal Prussian Cuirassier Regiment "Count Gessler" (Rhenish) No. 8), in which the uniform had medium green cap bands, collars, and cuffs. In addition, the service cap of the cuirassiers was traditionally white (Profile 4).

Since Hans Bethge as well as Gustav Nernst had the upper surface of the upper wing painted a particular color, this may have been the special marking for a Ketten leader.

Lt. Joachim von Bertrab had the fuselage of his Albatros D III painted black (Profile 5). The national insignia on the fuselage was white. This paint scheme as well was related to his original unit, the 46th Field Artillery Regiment, whose uniform had black cap bands and black collars.

Above: The airfield at Phalempin (Stiftung Deutsches Technikmuseum Berlin/Historisches Archiv)

D.2126/16, delivered to the Staffel on March 27, 1917, had as a personal marking a black "X" edged in white as a personal marking and was flown by Uffz. Josef Heiligers (Profile 6).

Paul Erbguth's Albatros D III D.2140/16 had as a personal marking a red "E" on both sides of the fuselage. The rudder was white with a black edge (Profile 7).

Lt. von Schell's D.2305/16 bore a fuselage band with the Prussian colors of white and black on the fuselage. The rudder was white with a black edge (Profile 8). The Albatros D III of Lt. Seitz had a red "S" as a personal marking on the fuselage. Here too the rudder was enclosed with a black edge (Profile 9).[14]

Lt. von der Marwitz, who was transferred to the Staffel in April 1917, had the fuselage of his Albatros D III painted wine red, while the engine cowling was light blue. This airplane too had the black bordered rudder. Wine red was his favorite color and the light blue was a gesture of respect to his original unit, Ulanen Regiment Hennings von Treffenfeld (Altmärkisches) Nr. 16, whose cap bands, collars, and cuffs were light blue (Profile 10). Because later almost all of the aircraft of Lt. von der Marwitz were painted wine red, Hans Holthusen often teased him with the remark that he had his aircraft painted the color of his favorite beverage (red wine).[15]

As is evident in the examples mentioned, a whole string of pilots chose the color of the cap bands of their original units for the personal paint scheme of their aircraft. In this connection one must not forget that even in their new function the pilots still wore the uniform of their original units and normally still felt strongly attached to them. They were detached to the flying service and were recognizable as pilots only by way of the flyer's badge on their chests. (The only exception consisted of the officers and ranks who joined the air service directly).

The black edged or white rudder may have been the Staffel marking. However, in May 1917, after the assignment of new Albatros D IIIs (from the D.600/17–D 649/17 and D.750/17–D.799/17 batches) and Albatros D Vs, this was not continued.

2. June 1917 – August 1917

Trench Battle in Artois

After the end of the Easter offensive and the associated subsiding of combat actions on the ground, as well as the shifting of the focus of battle to Flanders, activity in the air on both sides in the sector of the German 6th Army decreased.

On the evening of June 1, several machines of the Staffel were aloft under the leadership of Oblt. Bethge in the southern part of the sector of the neighboring 4th Army. North of Comines they came across some F.E.8s of 41 Sqn RFC and in the resultant air combat the Staffel commander was able to achieve his 7th aerial victory.

On the next day an order arrived at Jagdstaffel 30 which stated that the Staffel, together with Jagdstaffel 27, was to shift its combat activity exclusively to the Wytschaete salient in the northern part of their sector. However, all the operational flights in this sector remained without results.

On June 5, 1917 two new pilots arrived at the Staffel in the persons of Lt. Hans Schlieter and Lt. Otto Fuchs.

Hans Schlieter hailed from Lehmin near Potsdam, where he was born on April 3, 1893. He belonged to that group of pilots in the air service who had already completed their training before the war. Since 1914 he had been active with Feld-Flieger-Abteilungen 42, 67, and 15, as well as Flieger-Abteilung (A) 223 on the Western Front as well as in Russia before he had joined pursuit aviation and so already had a great deal of frontline experience.

Otto Fuchs was born on March 7, 1897 in Frankenthal in the Palatinate (at that time part of the Kingdom of Bavaria) as the son of a teacher. He was already fascinated by aviation as a child. When a teacher in primary school once asked him what he wanted to be when he grew up, he responded: "Aviator!" As a youth he constructed model gliders, which he enabled to take off by means of a rope mechanism. His largest model had a wingspan of two meters. He even constructed a towing device with which two aircraft attached together were pulled into the air. Once he even put his cat as a "pilot" in one of his aircraft, which it did not enjoy at all, and after the landing it gave him some scratches on his hands and face.

He furthermore harbored a great interest for literature, philosophy, and painting. For this reason, after his Abitur [secondary school final exam] he began to study philosophy in Heidelberg.

After enlisting voluntarily at 17 years of age, he went to the front with the Bavarian 12th Field Artillery Regiment on October 8, 1914. After his promotion to Leutnant, he joined the air service at the beginning of 1916 in order to become a pilot. In March 1916 he was briefly ordered to Feldflieger-Abteilung 18 in order to determine his suitability

Above: Five Albatros of Jagdstaffel 30 in June 1917. Second from the right is the D V D.1012/17 of Lt. Erbguth. On the far left one can see the machine of Lt. Seitz. The pilots of the remaining aircraft are not known.

Above: Four Albatros D IIIs and two Albatros D Vs stand ready for take-off on the Staffel's airfield at June 1917. The second airplane from the right has a black/white/black band around the fuselage as a personal marking. The fuselage of the aircraft is presumable painted in silver. (Stiftung Deutsches Technikmuseum Berlin/Historisches Archiv)

for the air service. The commander of the Abteilung assessed him as follows:

Leutnant der Reserve Otto Fuchs has flown with the Abteilung several times. According to the opinion of the pilots, he is quite especially suitable for deployment in the air service.[1]

On August 15, 1916 he was transferred to Bavarian FEA 1 at Schleissheim and began training as a pilot. After its successful completion he went to Flg.Abt.(A) 292 in February 1917. In April 1917 he reported to the fighter arm, for which reason his Abteilung commander Hptm. Franz Hailer composed the following assessment on April 20, 1917:

The officer has been with the Abteilung as a pilot since 13.2.1917 and during this time developed into an outstanding flyer. His enthusiasm for flying and his fresh daring had an energizing effect on the spirit of the other crews. I consider him quite especially suitable for becoming a fighter pilot. He is certain to have a successful career. His calm modest manner and his flawless military and social conduct combined with the above-mentioned merits in a worthy fashion and made him a popular and respected comrade.[2]

On May 29, 1917 his wish was fulfilled. He was transferred to Aflup 6 and ordered to the Jagdstaffelschule at Valenciennes.

When Otto Fuchs arrived a week later, on the morning of June 5, 1917, at the railway station of Phalempin and was picked up by Douglas Schnorr in the Staffel auto, his training had turned out completely differently than he had imagined. Already during the second practice flight in a single-seater fighter he had been cut off by another aircraft. In the attempt to avoid a collision his airplane "wiped out" in the direction of the flight instructors, who were just able to throw themselves to the ground. After the landing there followed a terrible row, in the course of which a furious flight instructor informed him that he had no further business at the fighter pilot's school and that he would have to leave. Otto withdrew completely stunned and almost did not notice that a red Albatros landed at the airfield, about which he reports:

Drawn by an uncertain presentiment, I strolled over there. A rotund puppet lifted himself out of the cockpit. Out of his furry wrapping peered a 'Pour le Mérite'... An „Kanone" [a Cannon - in the flyer's jargon of the day, the designation for an above average successful fighter pilot—the authors]. Would he take you? And already I was striding towards the smart little fellow, who in the meantime had removed his wrappings. It was Leutnant Allmenröder. I told him of my misfortune and asked him to request me for the Richthofen Staffel. The following afternoon he appeared in an automobile, had me demonstrate my flying a bit, and took me along.[3]

His stay with Jagdstaffel 11 was however very short, then he was ordered to Jagdstaffel

30. Oblt. Bethge had inquired after Otto Fuchs at the Jagdstaffelschule and discovered that Karl Allmenröder had taken him to Jagdstaffel 11. Bethge, afterall with the rank of Oberleutnant, immediately complained to Leutnant Karl Allmenröder that he had "snatched away" a pilot intended for his Staffel. As Otto Fuchs later remarked, this was the best thing that could have happened to him; with his modest flying abilities and his ambition at that time, he probably would not have long survived the "heroic air" of the Richthofen Staffel.

After Otto Fuchs arrived at Phalempin, Lt. Douglas Schnorr showed him his quarters. It was the room in which Gustav Nernst had lived beforehand. For this reason Douglas Schnorr asked him whether the room was all right with him or whether he wanted a different one. Otto Fuchs was a little surprised by this question and Schnorr explained to him:

It is Bethge's wish that I take the idiosyncrasies of the gentlemen into consideration. As Offizer z.b.V I am just a glorified nursemaid anyway. No one should feel cramped or violated here. If someone maintains that he can only sleep in a bed facing from east to west because otherwise he'll be plagued by the 'Od' or the Devil, then by God it'll be turned in that direction. You laugh? It's happened before.[4]

Otto Fuchs remarked in addition in a conversation that almost every pilot with whom he became acquainted trusted in some sort of good luck charms or rituals. The Scottish aviation historian Alex Imrie—himself a professional pilot—confirmed this, but added, "Hardly anyone will admit it..."

After he had gotten set up in his quarters, Lt. Schnorr brought him to Hans Bethge. Otto Fuchs had just begun to report about his experiences at the Jagdstaffelschule and his time with Jagdstaffel 11 when a call from the front arrived reporting the incursion of enemy aircraft. Hans Bethge jumped up, ordered the take-off of the Staffel, and asked Otto whether he would like to join the flight. He could use his reserve machine. When they arrived at the airfield, everything there was already in motion:

Clattering, roaring, shouting, and cursing. From the long row of colorfully painted Albatros fighters one, two separate themselves... stagger, gain speed, and lift off from the ground. On some of them the propellers are turning slowly like the sails of a windmill. On others they are whipping the air in a whizzing circle. Dust is flying and caps are swirling. Strong mechanics brace themselves against trembling wings as the wind tears at them. Where is Bethge? There he is, just now swinging into his cockpit. He looks aroung searchingly, waves at me, and then points to the left. In a few bounds I am standing by my airplane.

"How many revs?"

"Thirteen eighty. Everything's set... loaded."

I adjust myself in my seat. A mechanic bends over me and fastens the shoulder and body harness. Goggles down... gloves on... ready!

"Clear!"

"Clear," someone answers behind me.

Throttle forward! My bird leaps onto the field with a loud roar, rocks, hops, and then floats. I immediately go into a turn, pull on the stick energetically, and look below me over the wing where the rotating field sinks lower and lower. One machine flits past close by me. I run into its slipstream, am roughly tossed back and forth, and decide to pay more attention. Now Bethge is taxiing out. It looks really amusing the way the dainty toy runs faster and faster over the ground. Then its shadow detaches from it. He is flying. I rush to get close to him. But others are more agile.

Above: Oblt. Bethge on the Staffel airfield in a pose which, according to Otto Fuchs, was quite typical for him.

This Page: On the occasion of the visit of the Kofl of the 6th Army shortly before June 10, 1917, ten Albatros D IIIs and four Albatros D Vs stand arranged in parade formation on the airfield of Jagdstaffel 30 according to Ketten (Bethge – Erbguth – von Bertrab – von der Marwitz). (Middle and bottom photos Stiftung Deutsches Technikmuseum Berlin/ Historisches Archiv)

Right: On the occasion of the visit of the Kofl of the 6th Army shortly before June 10, 1917, ten Albatros D IIIs and four Albatros D Vs stand arranged in parade formation on the airfield of Jagdstaffel 30 according to Ketten (Bethge – Erbguth – von Bertrab – von der Marwitz).

From all sides they dash on by. They quickly align themselves behind the leader. I just manage to secure a spot at the end of the formation...[5]

The first fighter patrol was extraordinarily strenuous for Otto Fuchs, as the whole time he was preoccupied with not losing his comrades. Fortunately a wine red Albatros appeared next to him regularly, whose pilot apparently had made it his business to keep an eye on the "new guy." Otto Fuchs had soon completely lost the overall view of things and "stuck" to the red bird until landing. He noticed nothing of the aerial combat fought and the victory achieved during this fighter patrol. Only in the mess did he learn from his new comrades that Hans Bethghe had forced a Spad to land to the east of Frelinghien.

After his first operational flight as a fighter pilot Otto Fuchs was overcome by strong doubts as to whether he was cut out for this "business." On the next day he spoke about it with Hans-Georg von der Marwitz, in whom he had immediately placed his trust due to his engaging manner. Hans-Georg von der Marwitz laughed:

"You poor little novice, that happens to everyone when they aren't used to the high altitude. One simply falls asleep. Of the eight Englishmen that Bethge most recently attacked I detected a mere three. And Brügman, the high spirited little dragoon, asked me when Bethge reported his victory, where is that supposed to have been... Nice, eh?"

"And what is he doing about this state of affairs?"

"Nothing. He has faith in his teddy bear."

"Oh, stop it."

"Really! Why else would he have tied it to the strut? He inherited it from Gustav Nernst. The one time he left it home."[6]

Above: The Albatros D III of Oblt. Bethge is shoved back into the hangar at the end of the visit. (Stiftung Deutsches Technikmuseum Berlin/Historisches Archiv)

During the further course of the conversation Hans-Georg reassured him with the statement that according to his experiences one would get used to the great altitude in time.

In June 1917 the pilots were divided up tactically into four Ketten, which were led by the Staffelführer Oblt. Bethge, his acting replacement Lt. Erbguth, as well as Lt. von Bertrab and Lt. von der Marwitz. Otto Fuchs described the deployment of the flights as follows:

The division into Ketten played a role especially during the deployment of the entire Staffel. If only a part of the Staffel were deployed, then this would take place in one or two Ketten. To fly alone at the front one needed Hans Bethge's permission. The division into Ketten meant that when possible the flight was deployed in closed formation. Of course, this wasn't always the case. If the situation demanded swift action, or a pilot couldn't take off, a Kette was composed of pilots from various Ketten. During my second victory, of the two pilots who accompanied me, only Rudolf von der Horst belonged to my Kette. Beginning in the second half of October, Uffz. Liebert was the third man in my Kette.[7]

Otto Fuchs was detailed to the Kette of Hans-Georg von der Marwitz, to which Uffz. Josef Heiligers also belonged.

The pilots whom Otto Fuchs met in the Staffel became, with few exceptions, trustworthy comrades. He soon developed a friendship with some of them. This was the case especially with Hans-Georg von der Marwitz, who with all his carefree ways and occasional high spirits gave the pilots of his flight a feeling of security and reliability, and the always chipper "little dragoon" Heinrich Brügman. He had especial confidence in Paul Erbguth because of his calm and friendly manner.

The center of the pilots was the Staffel commander, Oblt. Hans Bethge. With his 27 years he was considered "old" as a fighter pilot and for his young pilots he represented a father figure whom they trusted almost blindly on the ground and above all in the air. Hans Bethge was a man of great courage, very good flying ability, and extraordinary leadership abilities.

Manfred von Richthofen, who visited him several times in April 1917 in order to coordinate the missions of the Jagdstaffeln in this sector of the front, had a very high opinion of him and his abilities as a Staffel commander. Under Hans Bethge Jagdstaffel 30 became a successful Staffel. At the same time the security of his pilots always preceded his personal success. His main concern was bringing young, inexperienced pilots home safely. After every operational flight he assembled his pilots for a thorough post-mortem and discussed the behavior of each individual, with indications as to what had to be improved in the future. He trained his new pilots tirelessly and demanded of them that they cover each other and carry out an attack only from the best tactical position.

"If you carry out an attack, look for the best position and consider whether the weakest man in your Kette can also carry out this attack safely," he urged Otto Fuchs when, after Lt. von Bertrab was taken prisoner in August 1917, he took over the leadership of his Kette.[8]

Otto Fuchs soon noticed that Hans Bethge was deeply affected after the loss of one of his pilots and quietly reproached himself with the question as to whether he had personally done everything to avoid the death of the subordinate: this was the brooding side of his nature. In time he became more and more disillusioned, he increasingly doubted the sense of the war and the goals of the political and military leadership. He naturally kept these thoughts to himself. Only Otto Fuchs learned about it in a personal conversation shortly before his transfer to another Staffel.

Above: Vzfw. Hans Oberländer. (A. Imrie)

Hans Bethge had a great interest in literature, philosophy, and art. He had a constantly growing library in his quarters and, when there was an opportunity, he went to French bookstores to look for further books. His favorite subject, as with Otto Fuchs, was the German Romantic movement and soon the two formed the "philosopher's club" of the Staffel.

In the same month as Otto Fuchs, 21-year-old Lt. Karl Weltz from Speyer in the Palatinate also arrived.

Above: Lt. Oskar Seitz in the pilot's seat of his newly taken over Albatros D III D.767/17. (R. Kastner)

Below: Lt. Oskar Seitz in front of his Albatros D III D.767/17 which in the meantime has been decorated with his personal marking. (R. Kastner)

Above: The black Albatros D III of Lt. Joachim von Bertrab in June 1917, since painted with the red shell edged in white. In the meantime the machine has been pushed from the parade formation to in front of the airplane hangar, where one of the Excellencies visiting the Staffel has the function and operation of the instruments explained to him by Hptm. Sorg. (A. Imrie)

He was an enthusiastic sportsman and began every morning with calisthenics and gymnastics. Already in January 1915 he had reported to the air service and acquired his first frontline experiences as a pilot in Kampfstaffel 4 of Kagohl 1. On April 25, 1917 he achieved his first victory by downing a B.E.12 near Doiran Lake in Macedonia. Because of the changing of this Kampfgeschwader into a Bombengeschwader with large three-seat aircraft shortly thereafter, he had reported to become a fighter pilot. According to statements by Otto Fuchs he was an outstanding pilot, but a lousy marksman.

The non-commissioned officers amongst the pilots formed their own group, although Hans Bethge made very sure that all pilots were treated alike. The non-commissioned pilots dined with the officers and also had access to the large living room of the villa that served as an "officers' mess." But there was also a mess for non-commissioned officers. Hans Oberländer was considered a spokesman of the non-commissioned officers whose word also carried weight with the officers.

When on the morning after some days of his arrival at the Staffel Otto Fuchs arrived at the airfield there was extraordinarily lively activity. The mechanics were shoving all available aircraft onto the field before the hangars and lined them up according to Ketten in parade formation. The reason for this "red carpet display" was the visit of the Kommandeur der Flieger of the 6th Army, Hauptmann Sorg, who was to appear in company with higher ranking officers in order to inspect the Staffel.

Otto Fuchs got hold of a ladder and climbed onto the roof of the aircraft hangar located on the southern part of the airfield in order to photograph the lined up airplanes. After the photos were developed, every pilot of course wanted to have prints of it.

Sunny early summer weather on June 7, 1917 led to increased enemy aerial activity. At the same time the Staffel undertook already in the morning several flights over the southernmost section of the front of the German 4th Army. At about 8 o'clock a Kette from the Staffel attacked a formation of F.E.2ds which had bombed the Château du Sart and were flying home. Uffz. Heiligers was able to force one of the F.E.s out of the formation and caused it to crash near Koelberg.

Three hours later the Staffel achieved a further success in an aerial combat with Sopwith single-seaters near Roulers, one of them being brought down by Vzfw. Hans Oberländer. In the afternoon aerial activity then slackened off distinctly due to a

Above: Lt. Joachim von Bertrab in his black Albatros D III. (R. Absmaier)

Right: Lt. Joachim von Bertrab. (R. Kastner)

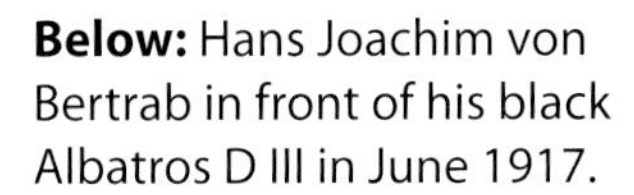

Below: Hans Joachim von Bertrab in front of his black Albatros D III in June 1917.

developing storm with isolated showers of hail.

On the next day Otto Fuchs experienced how quickly the situation in the air could change and how the attacker could become the hunted. The Staffel had taken off with several individual Ketten in order to support the units of the German 4th Army lying to the north. While doing so a Kette from the Staffel encountered an enemy formation, about which Otto Fuchs reported:

In the meantime we have approached one another closely enough to clearly distinguish friend from foe... the reported F.E.s slowly turn behind one another as though they were forming a circle. Above them, light and nimble, just as many Albatros fighters are performing gymnastics like playful little butterflies and attempt in vain to close in on the well-armed monsters. Now they seem to have discovered us. Their lead aircraft fires off several green flares. At that, all six raise their tails almost simultaneously in order to take their leave westwards.

The Germans dive after them. Unfortunately, we are much too far away to take part in the beginning chase. One Albatros especially distinguishes itself. Far ahead of its comrades, it fastens onto the heels of one lattice-tail that is diving headlong at an ever steeper angle. Then the five others unexpectedly turn about. At that moment the attacker finds himself in the middle of them and he in turn throws his machine around. Something white flutters away. His wings fold up and he plummets like a stone into the abyss... The Englishmen assume formation and fly at a low altitude back home.[9]

The Albatros shot down was Vfw. Franz Bucher, who had been with Jagdstaffel 30 for just five weeks. The first weeks were considered the most dangerous for newcomers to the front. The "green" flyers were still too much occupied with controlling the aircraft, keeping their orientation, or not losing their connection to the Staffel, as Otto Fuchs quite vividly describes in a report:

Near Armentières, the frequently winding ribbon of the Lys meanders through the network of trenches... Then I am suddenly startled. In front of us, about three hundred meters lower, two aircraft appear. They quickly approach and then flit past beneath us. I did not make out their insignia. Nor the type. Were they noticed by the others? In any event, it would not hurt for me to stick close to the formation and not lose sight of the two. But where are they hiding? I crane my neck around, lean to the right and left out of my machine, and search the dangerous region beneath my tail. They have vanished without a trace. How do they manage to do that anyway? By chance I cast a glance above me. Oh, they are up there! But they have doubled their number... No, they are altogether different machines. Where did those come from? Then suddenly little white clouds of shrapnel burst around them and immediately begin appearing by the dozen. They turn away and head in an easterly direction. The shooting stops. They were Germans.[10]

Above: Lt. Hans-Georg von der Marwitz with his dog "Duc" in front of the Staffel quarters. Like many other Staffeln, Jagdstaffel 30 also had a whole "pack" of dogs.

In another report Otto Fuchs describes the peculiar mixture of the charm and beauty of flying and the permanently present dangers during operational flights:

We swing back towards the front in a wide arc. Above our heads and below us there is a continuous coming and going of aircraft. I can make out some

Left: The Albatros D V D.1016/17 "red M" of Lt. von der Marwitz in June 1917. Behind it is a large Friedrichshafen G II or early G III bomber.

of them as Germans, but am completely uncertain about most of them. Somewhat farther back the artillery planes are describing their patient circles. Whenever we cut across the path of such a machine the observer puts on a fireworks display with star shells. He only calms down again when Bethge makes his own nationality known through the same signal.

Rrack... rrack... pomm... rrack. There is a sulfurous stench and a metallic clang. A dozen barking snowballs flutter around my machine like Chinese lanterns... and the rest of my formation?

Far to the right, almost a thousand meters lower. The realization of my negligence bores into my heart with glowing needles. I tear my machine upwards like a swallow, let it tip over its wings and dive with howling haste after my comrades.

While the disc of the earth reels in an incomprehensible curve I just manage to catch the shining triangle of Zillebeke Lake, and above that the pale stone wasteland of Ypres. I dive to the east of it as though every single-seater in the world were sitting on my neck. Once more I cross paths with a shrieking pack of shrapnel. Then I catch up with the Staffel. Only quite gradually do I manage to reassure myself regarding my carelessness. I make the best of resolutions, with the intent to never occupy myself with things which distract me—and nearly miss how Bethge turns to the north. No, I really am a bloody novice.

I notice that Bethge continuously flies straight ahead without paying attention to all the flying activity going on around us. He apparently wants to gain altitude. We have already reached 4200 meters. A severe cold makes itself noticeable which penetrates relentlessly through my fur coat and gloves and into my blood and brain.

Fifteen minutes later I read 5400 on the altimeter. The transformations below continue, tiny lights blink on the ground, little flames flash. West and east have flowed together into a single pool of bluish-black ink. For those down below night has fallen. The sun is taking its leave of us as well. At an incomprehensible height it begins to flatten out and shrivel up. It resembles an orange split right down the middle. One is tempted to believe that it simply slips into a slit in the ether. And suddenly it dawns on me: the azure green sky behind which it is hiding is the sea. If I look more closely I can even see the edge where it borders on the actual sky.

And what I had earlier thought to be the horizon, this thin flesh-colored strip, is the coastline. It stretches with fantastic purity from Holland to my feet, then to Cap Gris Nez, and from there leaps back towards the south. Yes, even the funnel-like mouth of the Cauche is visible and behind that the great glassy tube of the Somme. Like a fata morgana, nearby and at the same time as transparent as a breath of air, are shimmering the walls and towers of Nieuport. Close to that, darker and more venerable, are those of Furnes. From here a network of canals leads inland, running parallel to the dunes to the next large city: Dunkirk. And more and more distant places become visible which harbor new treasures, new cities sparkling like precious jewels. High in the background I can envision Abbeville.

Again and again I close my eyes and open them again, for what I am seeing does not seem possible to me, so inconceivable, so divinely beautiful is this image. With some displeasure I notice that Bethge is turning away. I follow reluctantly as the last man at a great distance. I am no longer making the slightest effort to maintain a particular position. Only now and again do I look ahead and am satisfied if I can discern the outlines of the nine birds.

Suddenly, just a few hundred meters ahead of me, a blinding torch flares up, shoots vertically into the depths, and is swallowed by the glowing jaws of hell. Right after that is another one... What was that?... With dismay I tear back the throttle and plunge into the abyss. The solitude which surrounds me is frightening. I imagine I can sense an invisible danger nearby and flee as though the Devil were behind me.[11]

On June 21st there was heavy mist, so the early patrol could not take off. About 7:30 the morning mist dispersed and very lively enemy aerial activity was reported from the front. Otto took off with one of the Staffel's Kette in a southerly direction, mainly along the rail line Lille—Douai in order to then turn in the direction of Arras. Southwest of Douai the three German Albatros encountered three English biplanes, concerning which Otto Fuchs reported:

...Now the three have once more turned towards the German lines. And appropriately the bursts also appear, though far to the side. It is clear that in the shimmering morning air nothing can be seen from the earth...

Now my clover leaf is separating. Two turn around and fly over the lines, while one maintains the northerly course. I chase after him and because I am faster than the fat biplane I slowly catch up. I am at exactly the same altitude. This is favorable. Diving a little, I will creep up under his tail and he will not be able to spot me. But why all the caution, since it is a German? I am already within a hundred meters of him. The observer is standing upright and doing some sightseeing. It is altogether certain that he does not notice me. He is not shooting. Of course. He has a clear conscience. He sees my crosses and is reassured. He is ranging German batteries and the flak can go to hell.

In order to no longer disturb the good Franz [observer], I turn around and am about to finally hurry homewards when I once more catch sight of the infuriating bursts of anti-aircraft fire. God knows, this time they are sitting in the midst of the two machines that turned away and which the recluse now joins – of course only to make himself independent again afterwards. This is just too strange! And what is yet stranger is that the two inseparable aircraft disappear westwards with their tails up while my friend still has not had enough. I also remember subsequently that I did not pay attention at all to the insignia earlier on. Nothing about the design of the aircraft strikes me as strange. From directly behind aircraft resemble one another just as people do. But it remains an unquestionable fact that the fellow did not open fire. No, he really did not shoot. One cannot fail to hear something like that. And then again, did he not stand upright and unsuspecting in his crate? He had a good conscience. A Tommy could never have that on the German side.

Above: Lt. Otto Fuchs at the beginning of his training as a pilot.

Amidst such considerations I creep up to firing range for the second time. The crate drifts calmly before me in the hazy blue. The Franz, probably bored, has settled comfortably in his seat. The bright apricot yellow leather cap moves at short intervals to the right and to the left. Again I occupy myself with the precise study of the silhouette before me. I compare all the known types, the English as well as the German. In vain. I do not find

Above: Otto Fuchs in front of his freshly painted "bilious green" Albatros D III. The national insignia on the fuselage and rudder surfaces have been over-painted.

any clues which would give me the right to press both of the buttons with which my nervous and impatient fingers are playing.

If only you would make a turn, you sluggard! If, instead of leaving your machine-gun sticking straight up in the air, you would just take a shot at me and reveal who you are... But there is no movement over there. The wood and fabric insect floats in front of my sights as leisurely as an ocean liner on a calm sea. Now I am less than 50 meters behind him. And if he still does not shoot now... Shoot, you lout! For heaven's sake, just shoot so that I can finally bang away!

He does not think of doing so. He sails along with a divine calm. A German... without doubt a German. Heaving a sigh, I reach for the throttle lever, in order to steal away sober and embarrassed.

At this moment the person in front of me makes a small turn. A tri-color rudder flashes in the glare of the sun. In the fuselage the figure of the observer jerks up, grabs for his machine-gun, and takes aim at me... My guns hammer away. Through the veil of powder I see the enemy go into a steep turn. I push the stick forwards in order to just scurry under his tail. The slipstream I cut across hits me roughly beneath my wings. I tumble, whirl, fall. During the attempt to turn for a renewed attack, I get the terrifying feeling that I am no longer capable of flying. How does one do it, anyway? A turn, how? Unbelievable! Have I completely lost my mind? Right aileron, stick into my belly, so... And again I have him in front of me. Admittedly somewhat obliquely... T-t-t-t-t-t... that was a salvo from over there. My guns reply.

Again I approach too closely. This time I tear my airplane closely above my opponent, turn, and get him once more into my field of fire. Then the monster rears up in front of me. Its propeller turns a few times feebly and then stops... Where are we? Trenches are writhing beneath me like worms. The sun? It is shining over there... We are on my side! Now just do not attempt to run away from me, otherwise... He tries it anyway and glides towards the west... Turn around right now! Tack-tack-tack... Will you not immediately give up every attempt to flee? Tack-tack-tack-tack-tack... what is he doing,

Above: Otto Fuchs next to his Albatros D III, the engine of which is just being run up by the mechanics.

anyway? The observer is lying with his upper body forwards on the fuselage. Something is going on. I have hit him seriously. Should I open fire once more? I cannot resist the temptation. I no longer let the enemy out of my sights and pursue him with clenched teeth. The observer with the apricot yellow leather coat is still lying half overboard. He is certainly dead... No, he is moving. He is raising his arm. Quite slowly he is raising his arm.

My vision goes black out of disgust and ferocity and I no longer release the button for the machine-guns. I fire and fire unceasingly, without aiming, driven solely by the desire to put an end to the spectacle before me, to completely destroy the remnants of the bleeding life, to get it out of my sight at any price.

Across the way a bright red spot slowly runs from the tip of the fuselage to the tail. The clumsy biplane tilts forward and races faster and faster towards the ground which turns beneath us deathly pale, red as meat, and disfigured by bluish spots. A small wood which has been shot to pieces stands out like the stubble of a beard, the shadowy profile of a railway embankment. During the dive his propeller begins to turn once more. For a moment the suspicion of a deception flares up within me. Again the little red mouse runs along the edges of the fuselage. It is no longer alone. Three, four, a dozen have joined it, a flickering streamer breaks

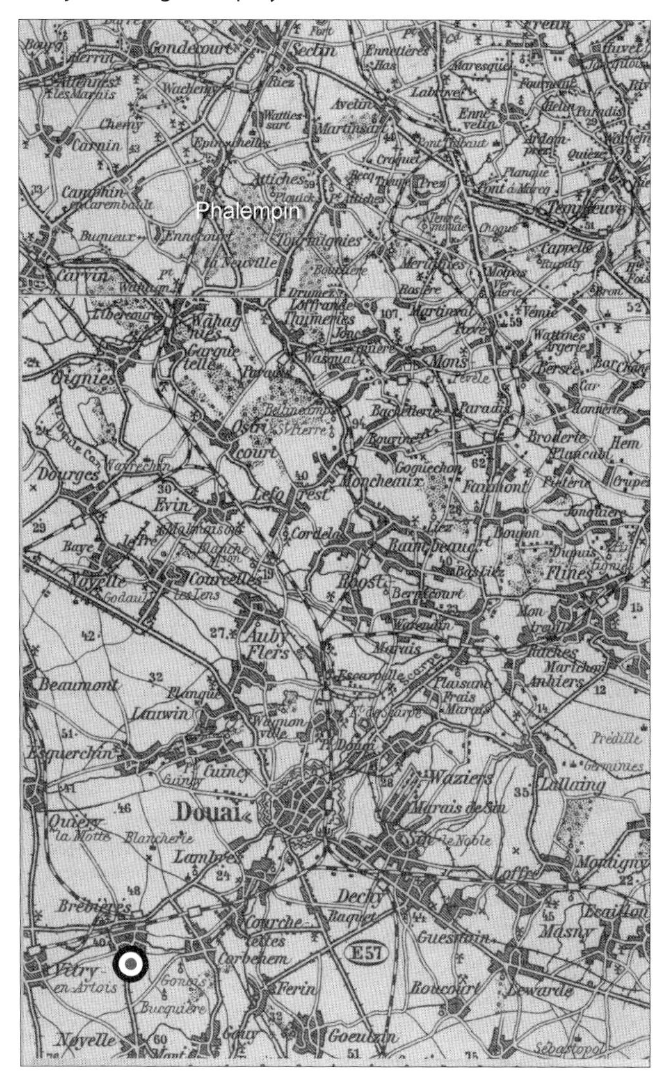

Right: The cockade marks the location of Lt. Otto Fuchs' victory over a Bristol fighter on June 21, 1917.

Above: Lt. Heinrich Brügman returns from a flight over the front.

out... Fire! A second later a black scorching blaze engulfs the plummeting foe. The moment he impacts the ground a towering column of flame shoots skywards. As a gust of wind parts the thick billowing clouds of smoke I see the glowing white skeleton silently collapse.

I am shaken by a sudden bout of shivering. Down below are the charred remains of my first victory! I am reeling. Good God, good God! I am sickened to the point of vomiting. My teeth are chattering from the cold. Cold? All the fires of hell are burning in my veins. I flee before horrible ideas and images which ensnare me.

Having arrived back home, I wander around with flushed cheeks unable to find any rest and am glad to make use of Bethge's offer to drive with the car to the crash site...

There curves the embankment. It has been bored through everywhere and viewed from up close it actually consists only of a host of mouse holes. The supporting troops of the infantry are lodged therein. We ask one of the ragged figures who follow us curiously and suspiciously with reddened eyes where the English aviator lies and have ourselves led over there by a grinning half-wild man. Three or four bearded, foul-smelling men of clay, tossing square tent cloths over their shoulders, join us.

There is little left anymore of the machine. A black frame of bent steel tubes, melted wires, an engine with its nose stuck in the mud... Everything burnt out and unusable. Nothing more is left over in such cases. One of the men standing around boasts in a childlike way that he had been a witness... After the crash he had immediately run over. One of the occupants had been flung out. "He was lying there..." Twenty paces away he indicates the spot. "When I came he was trying to stand up straight. He was just a bit daft..." He points at his forehead. "Otherwise nothing was wrong with him. The other one burnt up. They then brought the observer to Douai."

"The observer?" I do not understand.

But a sergeant from the battery dug in nearby confirms it. The observer had been taken prisoner almost uninjured.

Two days later Schnorr hands me the statement of the captured Englishman. His aircraft, a quite new type from the Bristol firm, was deployed for the first time on the front a week previously. Regarding the course of the combat, he states that until the first shots were fired he thought I was one of his comrades. The pilot was fatally hit right at the beginning of the fight. He, the observer, had tried to throw him out of the seat in order to steer himself. With bitterness and indignation he declares that I still fired long after he was completely defenseless. A hit in the tracer ammunition had caused the fire, from which he was saved only by some miracle.

Above: A "Kette" of Jagdstaffel 30 taking off for an operational patrol.

That is how my victory looked as viewed from the enemy side.[12]

To one of the authors, Otto Fuchs added this account:

"When I received this information from Douglas Schnorr, it was as though I were thunderstruck. Because of the crazy flight maneuvers, I had assumed that I had a crafty opponent in front of me. It did not enter my head that the observer was desperately trying to avoid a crash. I felt very sorry afterwards. I should have actually told that to the captured Englishman, but that would have seemed to me like an artificial belated justification.[13]

The victory was officially confirmed by the Kogenluft as his first victory. The surviving gunner was Sgt. W Mollison, who was taken captive unwounded. The pilot, 2/Lt. D C H MacBrayne, died.

June 24, 1917 brought lively enemy aerial activity despite heavy clouds and strong winds, which however was mainly limited to the airspace on the other side from the German front. In the evening around 7 o'clock there was an aerial combat with several Nieuports southwest of the railway station of Beaumont, of which Uffz. Heiligers was able to shoot down one after a lot of turning.

On July 2, 1917 it was once again shown how dangerous the first weeks at the front were for a newcomer. A Kette from the Staffel had taken off amidst relatively slight enemy aerial activity, when they were surprised by three Nieuport single-seater fighters around 11:30. Lt. Hans Forstmann, who had only been with the Staffel for two weeks, was shot down and crashed fatally by the canal near Dourges.

The opponents may have been Nieuports of 40 Squadron RFC, who reported the downing of two Albatros D IIIs by 2/Lt. Godfrey and 2/Lt. Crole in aerial combat north of Douai between 10:30 and 10:50, which the British pilots documented in their combat reports as follows.

Report of 2/Lt. Albert Earl Godfrey:

I was leading the flight at 13,000 feet over the front. I recognized an enemy aircraft north of Douai and climbed to 17,000 feet. Another enemy machine joined the one spotted. The flight turned and dived upon the enemy aircraft. My Nieuport dived on the enemy aircraft and I fired a salvo from about 25 yards distance. The engine of the enemy machine stalled and the aircraft crashed out of control.
I turned in order to look for the second enemy machine, but it had disappeared. I turned again and saw the first enemy aircraft still going down in a vertical dive.

Note: The enemy aircraft had red and green upper wing surfaces, as well as a fuselage band with white dots.[14]

Dienststelle : Kgl. Preuss. Jagdstaffel 30

(Standort) : Flughafen, den 30. Juni 1917

eingetr. f.

Personal-Bestandsnachweisung.

Veränderungen zur Kriegsrangliste. (Nur für Jdflieg. auszufüllen.)
Abgang : (wohin ?), Beförderung, Patentverleihung, Umschulung, Prüfung.
Bei Zugang : (woher ?) Kriegsrangliste auf besonderen Bogen beifügen.

(Unterschrift und Dienstgrad : Oberleutnant u. Staffelführer

Personal roster of Jagdstaffel 30, Juni 1917, part 1
Bayerisches Hauptstaatsarchiv Abt. 4, Iluft-Band 206

Flugzeugführer :

Lfde. Nr.	Dienstgrad	Name	Geburtsdatum	Patent	Truppenteil	Zugehœrig oder kommandiert	Abzeichen seit	Bei der Dienststelle seit	Woher zur Dienststelle gekommen	Bemerkungen
1	Oblt.	Bethge Hans	6.12.90	6.6.16	Eisenb. Regt. 4	Jasta 30	8.12.15	20.I.17	Jasta 1	Staffelführer
2	Lt. d. Res.	Schreiner Douglas	15.9.90	8.9.15	Funker Abtlg. 19	"	8.6.16	16.I.17	Fea 11	Offz. z. b. V.
3	Lt.	von Bertrab Joachim	18.X.94	22.III.15	Feld. Art. Regt. 100	"	Febr. 17	6.III.17	A. Fl. Sch. A. Straßburg	
4	Lt. d. Res.	Erbguth Paul	24.11.91	22.12.14	Landw. I. R. 107	"	9.II.17	8.III.17	Schusta 21	
5	Lt.	Brügmann Hermann	22.10.96	22.III.15	Drag. Regt. 7	"	28.10.16	28.III.17	Flg. Abtlg. A 240	
6	Lt. d. Res.	Seitz Oskar	22.I.92	12.III.15	5 bay. Chev. Leg. Regt.	"	19.4.17	14.4.17	" " " 292	
7	Lt.	v. d. Marwitz Hans Georg	7.8.93	22.III.15	Ul. Regt. 16	"	31.I.17	18.4.17	Jastaschl. O. K. L.	
8	Lt.	Schlieter Hans	3.4.93	23.6.12	Feld. Art. Regt. 17	"	10.12.14	5.6.17	"	
9	Lt. d. Res.	Fuchs Otto	7.3.1897	17.12.15	F. A. Regt. 12	"		7.6.17	A. Fl. Sch. 6	
10	Lt.	Forstmann Hans	9.6.94	22.3.15	I. R. 165	"	9.6.17	16.6.17	" " "	
11	Lt.	Weltz Karl	24.1.96	21.10.14	I. R. 27	"	2.12.15	25.6.17	R. G. 1	

Flugzeugführer

Lfde. Nr.	Dienstgrad	Name	Geburtsdatum	Patent	Truppenteil	Zugehœrig oder kommandiert	Abzeichen seit	Bei der Dienststelle seit	Woher zur Dienststelle gekommen	Bemerkungen
12	Vizefw.	Oberländer Hans	3.9.94	—	Fea 6	"	3.3.16	17.5.17	A. Flg. Sch. 6	
13	Uffz.	Heiligers Josef	23.4.94	—	Fea 7	"	18.8.16	20.I.17	Jasta 4	
14	Gefr.	Funk Josef	9.12.96	—	Fea 10	"	13.3.17	23.5.17	A. Fl. Sch. 6	

erku

Lfde. Nr. | Die gr

Personal roster of Jagdstaffel 30, Juni 1917, part 2
Bayerisches Hauptstaatsarchiv Abt. 4, Iluft-Band 206

Above: Spad VII A6747, forced to land by Oblt. Bethge on June 5, 1917. (A. Wait)

Above: F.E.2d A1957 of 25 Squadron RFC, forced to land on June 7, 1917 by Vzfw. Heiligers, is dismantled for transport.

Report of 2/Lt. Gerard Bruce Crole:

On patrol in the vicinity of Douai I noticed two enemy aircraft. I turned, attacked one of the machines and fired a salvo from point blank range. The right wing of the enemy machine broke off and the aircraft fell, going up in flames, simply downwards. While I circled, I looked for the other aircraft, but I could no longer see it.[15]

According to the reports of the two British pilots, both of the German fighter planes must have been completely surprised by their opponents. One of the German machines may have escaped during the dive, while the second with Lt. Forstmann was shot down. Such a surprise attack was the greatest danger for inexperienced pilots, who were still completely occupied with controlling the aircraft and orienting themselves.

If one reviews the month of June, the highpoints of which have been described above, as reflected in the statistics of the Kofl. 6 (for the exact period of June 1 to 28), then it will read fairly matter-of-factly as follows: the Staffel completed altogether 416 operational flights with a total of 500 flight hours. In the course of theses flights there were 141 aerial combats, of which five ended successfully with the downing of an enemy aircraft.[16]

As one sees, it was a considerable expenditure for an (apparently) meager result—so entirely different from one commonly imagines, where the flying heroes take off every day and land again rejoicing after umpteen victories. Most Staffeln had to "work hard" for their victories; only a few succeeded (and that mostly for a short period) in shooting down a series of opponents.

After the main battles had moved to Flanders, and therewith to the sector of the German 4th Army, the number of Jagdstaffeln in other sectors

Above: German soldiers inspect F.E.2d A1957, which was forced to land.

was "thinned out," in order to better face up to the enemy on the new main front. At the beginning of July six Jagdstaffeln were still being deployed, which number in the course of the month was reduced to two (Jagdstaffel 12 at Roucourt and Jagdstaffel 30 at Phalempin). The provision of Jagdstaffel 2 (which shortly thereafter was exchanged for Jagdstaffel 37) represented only feeble compensation. Now as before the enemy forces opposite these Staffeln were numerically superior.

Nonetheless, at first aerial activity decreased further. Thus Jagdstaffel 30 recorded in the three

Lt.d.R.S e i t z, Jagdstaffel 30 Phaleleinpin, 4.7.17.

Gesuch

Betr. Zurückversetzung zur bayer. Fliegerabteilung (A) 292

Ew. Hochwohlgeboren bitte ich gehorsamst meine Zurückversetzung zur bayer. Fliegerabteilung (A)292 genehmigen und höheren Orts befürworten zu wollen. Als Grund gebe ich Folgendes an: Am 21.4.17. stiess ich jenseits der Linie im Verlaufe eines Luftkampfes mit einer Alb.D III in 2700 m Höhe zusammen, wobei ein Teil des linken unteren Tragdecks, Fahrgestell und Propeller weggerissen wurden. Trotz der schweren Beschädigung des Flugzeuges kam ich bei Gavelle zwischen der deutschen und englischen Stellung noch heil herunter. Von den Engländern von den Trümmern meines Flugzeugs bis kurz vo die deutsche Stellung verfolgt, gelang es mir auch diese noch glücklich zu erreichen. Seit dieser Zeit besitze ich beim Fliegen innerhalb eines grösseren Geschwaders nicht mehr die nötige Sicherheit, um gleichzeitig mit diesem einen Luftkampf durchzuführen und hatte während meiner 2½ monatigen Tätigkeit keinerlei Erfolg. Nach Rücksprache mit meinem früheren Herrn Abteilungsführer kann ich als Flugzeugführer wieder zur bayer. Fliegerabteilung (A)292 zurück und ich glaube ausserhalb eines Geschwaderverbandes meiner Aufgabe als Flugzeugführer gewachsen zu sein.

S e i t z.

An Hochwohlgeboren
Herrn Oblt. u. Staffelführer
Jagdstaffel 30

B.B.Nr. Pers.102

Dem Kommandeur der Fl.6

Beistehendes Gesuch wird befürwortend weitergereicht.

Lt. S e i t z wird den Anforderungen einer Aufklärungs Fl.Abt. ohne Zweifel gewachsen sein. Bei einem solchen Verbande würden sicherlich noch wertvolle Leistungen von ihm zu erwarten sein.

gez. Unterschrift.
Oblt. u. Staffelführer.

Oberkdo. der 6. Armee
Kdeur d.Fl.
Nr. 790 pers.
Abt. IIa

An den kommandierenden Gen.d.Luftstreitkräfte mit der Bitte, Lt. S e i t z zum A.Fl.P. 6 zu versetzen.

A.H.Qu., 10.7.17.
J.V. gez. F r e y t a g.

Above: Lt. Oskar Seitz's plea to be transferred back to his old Fl. Abt. (A) 292 (Bayerisches Hauptstaatsarchiv, München, OP 12233)

Above: Lt. Seitz, Lt. Fuchs, and Vzfw. Oberländer inspect the remains of Bristol Fighter A7139 of 11 Sqn. RFC shot down by Otto Fuchs on June 21, 1917.

Above: The burnt-out Bristol Fighter A7139. In view of these charred remains, it is more than astonishing that the machine-gunner, Sgt. Mollison, escaped unharmed.

Above: The wreck of the Bristol Fighter A7139.

Above: Otto Fuchs and Oskar Seitz on the way home after they visit the wreck of the crashed Bristol Fighters A 7139. Otto Fuchs carries parts of the unburned fabric of the airplane as souvenier.

weeks up to July 19th 210 operational flights (231 flight hours) with merely 51 aerial combats Only two of these were rewarded with a victory.

On July 13th Uffz. Heiligers forced a Sopwith to land in the English lines near Annequin late in the evening. This success was to be later confirmed by the commanding general of the air service merely as "forced to land on the other side of the lines" and not as a full-value victory with a victory number. On the part of the headquarters providing recognition, one probably assumed that the destruction of the machine forced to land had not been proven beyond a doubt.

On July 14, 1917 six Martinsydes of 27 Squadron RFC flew over the front in order to bomb the airfield of Brayelle, near Douai. On the return flight they were attacked by Ketten from Jagdstaffel 3 and 30. One Martinsyde was shot down by a pilot of Jagdstaffel 3 at 7:50. A few minutes later Uffz. Heiligers forced a second Martinsyde to land between Leforest and Evin-Malmaison. The British pilot was taken prisoner unwounded. The remaining four British aircraft escaped in the direction of Avion.

On the evening of the next day there was an aerial combat between eight Albatros of the Staffel and four Sopwith Camels described as "Martinsydes" from 8 Squadron RNAS. During it Oblt. Hans Bethge was able to place himself behind a "Martinsyde" and cause it to crash in flames.

After a period of bad weather, in which aerial activity was reduced to a minimum, the Staffel attacked Sopwith single-seaters over British territory near Méricourt on July 20, 1917, whereby Oblt. Hans Bethge shot down an opponent at 21:25.

At this time the Staffel returned the Albatros D Vs allotted to it from the production batch

Above: The wreckage of Nieuport 23 B1607 of 29 Squadron RFC shot down by Vzfw. Heiligers. (R. Absmeier)

D.1000/17 to D.1049/17 to Aflup. 6 in accordance with Kogenluft. order No. 82632 Fl.III due to wing breakages appearing at other Staffeln in the beginning of July.

About two weeks previously, certainly after struggling with himself, Oskar Seitz had written the following request to higher quarters:

I obediently ask the Right Honorable Sir to grant my transfer back to Bavarian Fliegerabteilung (A) 292 and to recommend it to higher quarters. I submit the following reason: On April 21, 1917 I collided with an Albatros D III at 2700 meters altitude in the course of an aerial combat on the other side of the lines, whereby a part of the left lower wing, undercarriage, and propeller were torn away In spite of the severe damage to the airplane I still came down safely near Gavrelle between the German and English lines. Pursued by the English from the wreckage of my aircraft until shortly before the German line, I also succeeded in reaching it safely. Since this time I no longer possess the necessary self-assurance while flying in a large formation in order to carry out an aerial combat with the same and during my 2½ months activity had no sort of success. After consultation with my previous unit commander, I can return once more to Bavarian Fliegerabteilung (A) 292 and believe that I will be up to my duties as a pilot outside of a formation.[17]

The request was agreed to at higher quarters and so he was transferred back to his old Flg.Abt.(A) 292 on July 20, 1917. For this reason, Oblt. Bethge wrote the following assessment:

Leutnant d. R. Seitz has flown quite eagerly with the Staffel, however is no longer up to the demands of formation flying and aerial combat since on April 21 of this year he collided with a comrade flying in formation at 2500 (sic) meters height and only with effort escaped with his life. I believe that Lt.d.R. Seitz will be up to all the demands of a two-seater unit.

In the performance of his official duties Lt.d.R. Seitz distinguished himself through his high consciousness of duty and his good military education.

In the officer corps he is popular as a well-educated and always amiable comrade of exemplary modesty.[18]

In the period of his service with Jagdstaffel 30 Oskar Seitz had completed a total of 79 offensive patrols with a total of 78 hours flying time.

In July 1917 one mission almost ended fatally for Otto Fuchs. Driven by the ambition of following up his first victory with as many others as possible as quickly as possible, he attacked three British aircraft over the front with two Staffel comrades.

Above: Pilots of the Staffel wait for orders to take off on an August day in 1917. From left: Lt. Rudolf von der Horst, Lt. Hans-Georg von der Marwitz, Lt. Karl Weltz, Lt. Otto Fuchs, Lt. Paul Erbguth, and Lt. Kurt Katzenstein.

While doing so, he was left in the lurch by his two companions, a non-commissioned officer and an officer. Both had simply turned away and flown home. With numerous hits in his machine, he barely succeeded in escaping and making an emergency landing. Although he did not report this incident to Oblt. Bethge, the latter found out about it. The non-com was immediately transferred, and the officer followed a short time later. Leaving a comrade in the lurch was for Hans Bethge, who otherwise showed understanding for all his subordinate's problems, was an inexcusable offense.

Finally, Hans Bethge spoke with Otto Fuchs and urged him: *"... and do me the favor of reining in your ambition a bit."*[19]

But life in the Jagdstaffel did not just consist of flying, fighting, and dangers, but rather also had its pleasant aspects, as Otto Fuchs describes:

An extensive garden is part of the lordly residence in which we are accommodated like princes. Besides numerous green houses overflowing with colors and fragrances, it contains a lot of wall fruit, pears, plums, peach trees, and strawberry patches of enormous size and inexhaustible fertility.

If no particularly thick formations of aircraft are reported or if the weather does not prevent it, one can find us each day in the garden between two and four, widely strewn amongst the fragrant beds. In many especially bountiful spots one perhaps sees several squatting together, eagerly rummaging amongst the leaves and shoving one bit of fruit after another into their mouths amidst jokes and laughter. It presents a quite amusing picture when the Staffel commander and officers, one like the other, sit in the bushes up to their hips in their shirtsleeves and devote themselves to their pleasant activity without much movement. Once one has filled oneself here, one strolls in a leisurely manner to the airfield, where coffee is taken in a daintily furnished summer cottage. One rocks in creaking wicker chairs, smokes a cigarette, or stretches one's legs on comfortable deck-chairs, pages through a book, and listens to the indefatigable gramophone.[20]

The relations amongst the pilots was in Jagdstaffel 30, as in most other Jagdstaffeln, not very military, as the following episode shows:

Before the little terrace, whose red and yellow striped awnings fend off the oppressive July heat, is a babbling fountain. Weltz, wearing red triangular swimming trunks, stretches himself out there with a wet towel wrapped around his forehead and bakes in the sun. His chest and arms, a well-molded landscape of muscles, have taken on the color of tea. The monotonous melody of the water and the glaring afternoon light have lulled him to sleep. Erbguth, the Saxon, and Brügman, who is always in the mood for loose pranks, creep with buckets full

Above: Lt. Otto Fuchs in his new Albatros D V D.2140/17. (R. Absmeier)

of water and yank the unsuspecting victim out of his peaceful dreams with a hefty double gush from front and behind. Weltz, who gasps for breath like a fish, starts upright and gets a hold of Erbguth, who amidst his laughter is incapable of moving. Despite his pitiful cries for help he dumps him in the basin of the fountain in full uniform. However, Brügman, who is altogether satisfied with this turn of events, dumps the freshly filled pail of his mischievousness over the both of them.

"No, that... that is downright mean..." the betrayed Saxon stutters and wriggles like a trout in Weltz' merciless hands. Suddenly the latter lets him loose and throws himself with a single bound upon the dragoon, who adroitly hurls the bucket in front of his feet. Weltz stumbles and rolls in the grass. Before he regains his feet, the other is sitting in his aircraft, gets the propeller turning, and hops away across the field. Weltz, in no way lazy, swings into his single-seater as he is, in swimming trunks and dripping with water, and chases after him. For a quarter of an hour they tussle around several hundred meters above our heads, fire star shells in front of each other's noses, and buffet one another with the slipstream of their propellers. Just watching them fills one with fear and anxiety. Weltz reveals himself to be a superior flyer. He fastens onto Brügman so skillfully from behind that the latter, no matter how he twists and turns, is no longer able to shake him off and finally lands. The former flies one more steep round of honor in an impudent way and then sets his bird down right beside him.

"Hardly an afternoon goes by without something of this sort taking place. Bethge is not displeased to see a bit of mischief being carried out."[21]

Even the Staffelführer was not spared such mischief: Hans-Georg von der Marwitz had challenged his Staffelführer in some fashion and so Hans Bethge ran after him armed with a flare pistol. The hunt led as in a cinematic film over the roofs of the airplane hangars across rooms and the kitchen, through gardens and over fences. Finally Hans-Georg von der Marwitz reached the airfield, called to his mechanics, and took off with the first machine that was ready. Hans Bethge took off in the next ready machine and shortly thereafter both tussled about in an "aerial combat" about 50 meters over the airfield. After the landing they drank copious amounts of champagne to celebrate their reconciliation.[22]

Above & Below: Preparation for a test flight of Lt. Fuchs' Albatros D V D. 2140/17, which was repaired after the emergency landing caused by anti-aircraft fire. Traces of the hit by flak can still be seen on the nose of the aircraft.

Above: Lt. Otto Fuchs ready to take off in his Albatros D V D.2140/17. The aircraft was badly damaged at the crash during the "thunderstorm-flight" end of August (R. Absmeier)

Amongst the nicest of occurrences were the regular visits of Ilse von der Marwitz, the sister of Hans-Georg, who worked as a volunteer in a military hospital in Lille. She always came in the company of female colleagues who were glad to leave the horrors of the hospital behind them for a few hours. Such a visit meant that each pilot looked especially "chic" in order to make a good impression on the ladies, especially when Hans-Georg's sister was accompanied by a Miss Kuhlenkamp, with whom half the pilots had "fallen in love".

Douglas Schnorr then provided coffee and cakes and Paul Erbguth switched on his gramophone. Of course the young ladies were also regularly invited to go for a flight around the airfield in the Staffel's "hack." Ilse von der Marwitz on the other hand insisted upon flying as a passenger on such a trip around the airfield even in her brother's wine red Albatros D V.

But this idyll could change in minutes, then the pilots took off for their strenuous and dangerous mission. In the summer within a few minutes they had to put up with temperature variations of 25 to 30 degrees Celsius plus to sub-zero temperatures and in winter endure icy cold. The thin air at

Statt jeder besondern Anzeige.

Am 14. August fiel im Luftkampf unser unvergeßlicher Sohn, unser lieber, guter Bruder, Enkel und Neffe

Heinz Brügman

Leutnant im Westf. Drag.-Regt. Nr. 7
kommandiert zu einer Jagdstaffel

im Alter von 20 Jahren.

Die vorläufige Beisetzung erfolgte auf einem Ehrenfriedhofe in Feindesland.

Saarburg, Bez. Trier, den 19. August 1917.

In tiefer Trauer:

Landrat Dr. Brügman und Frau
Paula geb. Primavesi
Hertha Brügman
Carl Brügman, Leutnant i. Westf. Drag.-Regt. Nr. 7, z. Zt. im Felde
Maria Brügman
Kurt Brügman.

Es wird gebeten, von Beileidsbesuchen abzusehen.
(1b

Above: Obituary notice of Lt. Heinrich Brügman from 19th August 1917. (Szigeti)

Above: Pilots of the Staffel in August 1917 in front of the "readiness hut". From the left: Vzfw. Max Willmann, Gefr. Josef Funk, (guest), (guest), Oblt. Hans Bethge, Lt. Karl Weltz, Lt. Kurt Katzenstein, Lt. Paul Erbguth, Lt. Douglas Schnorr, Lt. Hans-Georg von der Marwitz, Vzfw. Hans Oberländer. The two rudders on the roofs belonged to an F. E. 2d of 20 Sqn. RFC (A6446, right) and a Nieuport of 29 Sqn. RFC (B1607, left). (Erbguth family)

higher altitudes caused problems. Colds, coughs, or earaches were the usual illnesses of the pilots. In addition, missions required a constant vigilance which led to a high degree of nervous strain. Many pilots were physically and psychologically exhausted and had to be ordered back home.

In the second half of July Otto Fuchs received Albatros D V D.2140/17, which had been transferred to the Staffel on July 12, 1917. However, already on one of his first operational flights with this airplane he had a rather unpleasant experience:

"On the flight back we spotted English trenches. Since we were relatively low, I went further down and fired upon the Englishmen from a low altitude. Suddenly I felt a violent blow to my machine and knew that I must have been hit by flak. My Albatros tilted down and nosed over, parts of the propeller splintered. Somehow I was just able to level out the aircraft. Fired upon heavily, I escaped back over the German lines in a glide and landed in a meadow. There I saw the mess: My motor had several hits from machine-gun fire, through which the mounting of the engine was torn, which finally led to my shooting a part of the propeller away. Shortly thereafter Heinrich Brügman landed next to me, jumped out of his Albatros, and ran towards me with the words, 'For God's sake, I was thinking you're bound to crash.' Upon viewing the damage, I realized just how narrowly I had missed crashing.[23]

At the beginning of the month of August aerial activity in Jagdstaffel 30's area of operations nearly came to a standstill due to bad weather. In the period of August 3 to August 9, 1917 only 19 sorties were flown, during which there was not a single aerial combat.

In this quiet time twenty-two year-old Lt. Kurt Katzenstein arrived at the Staffel, on August 9, 1917. He was born on February 27, 1895 in Kassel and came from an upper middle class Jewish family. He reported as a non-commissioned officer to the air service in July 1916. After his training he was transferred to Flieger-Abteilung 4, where on June 6, 1917 he was awarded the Iron Cross 1st Class.

After his application to be trained as a fighter pilot

Above: Lt. Douglas Schnorr in the pilot's seat of the Halberstadt D V "arranged" for him; the guns have been removed.

was accepted, he went to the single-seater fighter school at Paderborn. Afterwards he was ordered to the Jagdstaffelschule and on July 30, 1917 he was promoted to Leutnant der Reserve. Otto Fuchs described him as a young man with a very good education and exceedingly good manners. Courteous, obliging, but also willing to participate in every sort of practical joke, he quickly became a popular member of the Staffel. He soon joined in as the third member of the philosophy club, contributing the ideas of Jewish philosophy to its evening rounds of talks.

Kurt Katzenstein was assigned to Lt. Erbguth's Kette, to which Lt. Brügman also belonged. Already on the next day he took off with Albatros D III D.799/17 at 8:05 for his first operational flight with his Kette. There followed two more flying missions with Albatros D III D.799/17 on the same day, the last at 21:20.[24]

On the next day he undertook his first test flight with Albatros D V D.2150/17 and flew some circles around the field. But he carried out his subsequent operational flights with Albatros D III D.799/17.[25]

On August 12, 1917 Lt. Hans-Joachim von Bertrab took off for the front alone in his black Albatros D V around 15:30 with a cloudy sky and isolated showers. His goal was a British captive balloon near Souchez (southwest of Lens) that he wanted to attack, although lively enemy aerial activity was reported in this sector of the front. He did not return from this mission. Around 16:00 the report arrived from the front that an Albatros was attacked by enemy fighters and forced to land on the other side of the lines near Farbus Wood, about 4 km. south of Souchez. During his attempted attack on the balloon Hans-Joachim von Bertrab had been attacked by 2/Lt. Edward Mannock of 40 Squadron, RFC, about which the latter reported:

Had a splendid fight with a single seater Albatross Scout last week on our side of the lines & got him down. He proved to be Lieut. von Bartrap (sic), Iron Cross, and had been flying for eighteen months. He came over for one of our balloons—near Neuville-St. Vaast—and I cut him off going back. He didn't get the balloon either. The scrap took place at two

Above: A somewhat unsuccessful landing by Lt. Schnorr on the Staffel's airfield.

thousand ft. up, well within view of the whole front. And the cheers! It took me 5 minutes to get him to go down, & I had to shoot him before he would land. I was very pleased that I did not kill him. Right arm broken by a bullet, left arm, left leg deep flesh wounds. His machine—a beauty, just issued (June 1/17) with a 220 H.P. Mercedes engine, all black with crosses picked out in white lines—turned over on landing and was damaged.[26]

On the evening of the same day, during a combined deployment of Jagdstaffeln 30 and 37, there was an aerial combat with several British Nieuport single-seaters. Uffz. Heiligers succeeded in forcing one of the British pilots to separate from his formation and in an aerial combat to drive him down to the ground. Lt. Udet (Jagdstaffel 37) attacked this Nieuport, which finally made an emergency landing in the German lines near Courrières. Both German flyers laid claim to the victory, which in the end was granted to Josef Heiligers by a court of arbitration.

On August 14, 1917 Ketten from Jagdstaffeln 30 and 37 combined attacked a formation of Martinsydes. Lt. Heinrich Brügman fired on a machine also under attack by Lt. Udet, which shortly thereafter exploded in mid-air. When it did so pieces of the engine penetrated the upper wing of Brügman's Albatros, who was, however, able to return to Phalempin unharmed.

On the next day the Staffel's machines took off at 10:25 for an operational flight after lively enemy aerial activity had been reported. Included was the Kette of Lt. Paul Erbguth, Lt. Heinrich Brügman, and Lt. Kurt Katzenstein. At first the flight proceeded uneventfully. The Staffel was already on the return flight when it spotted what was identified as a formation of B.E. bombers in the vicinity of Hulluch, about 10 km to the west of Phalempin, and attacked it.

While doing so, Lt. Heinrich Brügman was badly wounded by a hit in the abdomen. He still succeeded in landing his airplane near Douvrin before losing consciousness. He was brought in by German soldiers and received makeshift treatment at the nearest dressing station. However, he died around 14.00 during the transport to a rear area military hospital.

Kurt Katzenstein had been separated from his Kette in the course of the combat and forced towards the northeast. Since he hardly knew the area and his gas was running low, he landed at the next best airfield (Wasquehal, Jasta 28's field) and after a 40 minute stay while fueling up, returned to Phalempin.

Lt. Brügman's opponents may have been D.H.4s of 25 Squadron, RFC. These reported an aerial combat

Besitz-Zeugnis.

Das laut Verfügung des kommandierenden Generals
Fl. I Nr. 106167 vom 6. Juni 1917 verliehene

„Abzeichen für Militär-Flugzeug-Führer"

ist dem
Leutnant Frhr. v. d. Horst (Rudolf), A.Fl.P. 1,
kdrt. z. Flieger-Abteilung (A) 267

ausgehändigt worden.

Hierüber wird dieser Beglaubigungsschein ausgestellt.

Im Felde, den 12. Juni 1917.

[signature]
Kommandeur der Flieger
der 1. Armee.

Above: Document "Badge for military aviators" of Lt. Rudolf von der Horst.

with 12 Albatros, of which one was shot down in flames by the crew of Capt. Morris and Lt. Burgess near La Bassée. La Bassée lies about three km northwest of Douvrin.

On this day, which was difficult for him, Otto Fuchs later remembered:

I had taken off with my Kette and came back around noon. In the mess we met only one pilot. From him we learned that Heinrich Brügman had made an emergency landing with a bad hit in the abdomen. The other pilots had driven to the hospital about an hour before in order to visit him. They returned in the late afternoon and went silently to their rooms. Hans Bethge looked in on us. I looked at him questioningly. He nodded briefly, shook my hand, and then likewise withdrew. The few who had remained behind in the mess spent the rest of the day in dismal silence.[27]

Towards evening Hans Bethge left his quarters and went to the airfield. Without informing the other pilots, he took off alone for a flight at the front. During this patrol he shot down a Nieuport at 20:45 over German-held territory north of Loison.

On August 16th there was an aerial combat between the Kette of Kurt Katzenstein and Sopwith Camels between Lens and Arras, which however resulted in no losses to either side.[28]

August 17, 1917 was a day with good weather, which led to lively enemy aerial activity over the front. In the evening one of the Staffel's Kette encountered a flight of Sopwith single-seaters about 10 km to the west of Phalempin airfield. Oblt. Bethge succeeded in shooting down a Sopwith near Wingles at 19:05. The crashing British aircraft rammed a second Sopwith, which likewise fell apart as it descended.

On the following day aerial activity was limited to the morning due to the ever worsening weather beginning at mid-day. A formation of eight Bristol F2bs of 11 Squadron, RFC was attacked around 7:50 southwest of Douai by flights from Jagdstaffeln 5, 12, and 30. The formation was split apart and two Bristol F2bs were shot down. Four of the remaining six Bristol Fighters tried to get away over Douai, but were pursued and attacked anew by an arriving flight from Jagdstaffel 12, whereby a further Bristol Fighter was shot down. Uffz. Heiligers reported the downing of Bristol F2b, but this machine was credited to Lt. Viktor Schobinger of Jagdstaffel 12.

At 9:00 the flight took off with Kurt Katzenstein and attacked two aircraft identified as "B E two-seaters" near Lens. Lt. Katzenstein reported the downing of a B.E., but the victory was not confirmed.[29]

There were also numerous aerial combats within the German 6th Army area. A formation of Martinsydes of 27 Squadron RFC were attacked on the return flight from a bombing raid on Aulnoye by Ketten from Jagdstaffeln 28, 30, and 37. In the course of the aerial combat Gefr. Josef Funk, together with Gefr. Friedrich Gille (Jagdstaffel 12) and Lt. Ernst Udet (Jagdstaffel 37), was able to shoot down a Nieuport between Bauvin and Meurchin. Since all three German flyers taking part put in claims for the victory, a court of arbitration had to clear up the facts, and finally granted the downed Nieuport to Josef Funk. It was his second victory.

On the same day Rudolf Freiherr von der Horst zu Hollwinkel arrived at the Staffel. He came from an old line of nobility which had produced the Secretary of State and a series of high-ranking officers. At the behest of his family he had joined a Guards Cavalry regiment, although, according to Otto Fuchs, everything military did not suit him. He was a quiet, sensitive, and delicate man who was interested in philosophy and poetry, wrote poems, with whom one could philosophize for hours. It is no surprise that he joined the Staffel's "philosopher's club." After completing his pilot's training with FEA 4, he gained his initial frontline experiences with Flg. Abt. (A) 267, where he was awarded the pilot's badge on June 6, 1917.

In the last week of August 1917 there was little aerial activity due to the bad weather. For this reason, Otto Fuchs made use of an afternoon free from rain and storm for an orientation flight in the area behind the front with the pilots newly transferred to the Staffel, Lt. Rudolf von der Horst and Lt. Kurt Katzenstein, during which he flew his "red F". He reported regarding this orientation flight:

When we had reached flight altitude it became more and more unpleasant. Heavy gusts of wind and an ever darkening sky heralded a strong summer storm. I became concerned about my two inexperienced pilots and decided to leave the altitude of about 3000 meters and to return to Phalempin in a descending flight.

Suddenly the black layer of clouds tore open, rays of sunlight shortly appeared, and I spotted a group of English fighter planes above us, who were apparently likewise on the way home due to the storm. From their superior position the Englishmen immediately went over to the attack and dove down upon us. Since I was flying higher than the two novices as protection, I was the first to be attacked by two Englishmen, and immediately let myself spin down.

While the bank of clouds approached threateningly, I was forced down lower and lower by my pursuers. I attempted to evade my pursuers while under constant fire, finally at the height of the tree tops. Again and again strong gusts of wind shifted my Albatros over several meters, so that I was afraid of ramming into a tree or a building. My machine had already been hit several times, when suddenly a hazy wall of rain appeared before me.

This was my salvation: I flew right into the wall of rain and immediately pulled the stick upwards in order not to crash into a tower or a tree. In the heavy rain shower I spotted a meadow below me and began to land. While doing so, my Albatros was pretty well damaged.[30]

When Otto Fuchs returned to Phalempin, soaked through and covered in mud, he discovered to his relief that his two comrades had also survived the air combat unharmed. Kurt Katzenstein had landed smoothly on the airfield with a few hits in his machine, while Rudolf von der Horst had been forced to make an emergency landing in his Albatros D V with a shot-up engine. After this experience, Otto Fuchs decided never again to use a letter as his personal marking.

Markings and Color Schemes of the Aircraft

In June 1917 the Staffel was equipped with a mixed complement of Albatros D IIIs and D Vs. In addition, there was a Halberstadt D V on hand used by adjutant Lt. Schnorr for flights in the area behind the front, but this was a mustered out school machine which Hans Bethge had "secretly" gotten hold of for his Offizier z.b.V. The guns of this "surplus" airplane had been removed and a white "12" was painted on the fuselage as a marking.

According to Otto Fuchs, there was no uniform Staffel marking at this time. Every pilot could paint his aircraft as he wished, as long as the basic color desired was not already taken by one of his Staffel comrades.[31]

The Staffel commander had retained the original blue-and-white marking on the rudder of his Albatros D III, adding a grey border, and additionally had a Mercedes star in blue and grey painted on both sides of the fuselage, since he was enthusiastic about the performance of his airplane's Mercedes engine (Profile 11).

Vzfw. Oberländer's Albatros D III bore the red letter "O" and a Feldwebel (sergeant) chevron on both sides of the fuselage (Profile 12). He was proud

Above: Visit by some ladies to the Staffel in the summer of 1917. In front, from the left: Ilse von der Marwitz, Hans-Georg von der Marwitz, Frl. Kuhlenkamp. In the rear, from left: Paul Erbguth, Karl Weltz, Douglas Schnorr, Rudolf von der Horst, Kurt Katzenstein.

to be a pilot from the ranks of non-commissioned officers.

The Albatros D III D. 767/17 of Bavarian officer Lt. Oskar Seitz had the red letter "S" on the fuselage and the rudder was painted with white and blue "Rauten" (lotzenges), representing the Bavarian Flag (Profile 13).

Lt. Heinrich Brügman had painted the fuselage of his Albatros D III D.2054/16 yellow and put the red letters "Br." on both sides. The reason for this paint scheme is not known (Profile 14).

The fuselage of Paul Erbguth's Albatros D V D.1042/17 was painted green, the tail was white, the horizontal tail surfaces green. It therewith represented the colors of the Kingdom of Saxony. On both sides of the fuselage the letter "E" was applied in red (Profile 15). Later he added a black outline to the rudder. He had the lower wings strengthened according to his own calculations, and was one of the few pilots of the Staffel who flew the Albatros D V without uneasiness (Profile 16).

One Albatros D V had the fuselage in silver gray and a black-and-white band around the fuselage. Lt. Forstmann was most probably shot down in this aircraft in aerial combat on July 2, 1917 (Profile 17).

The Albatros D III of Lt. Hans-Joachim von Bertrab was as ever painted black. However, he had a red star with tail and white edge painted on the fuselage. This insignia symbolized an exploding artillery shell. The national insignia on the fuselage and rudder were painted white (Profile 18). His Albatros D V had a black fuselage and the crosses were only outlined white (Profile 19).

Lt. von der Marwitz had a red "M" on both sides of his new Albatros D V D.1016/17. The tail of the Albatros D V and the metal parts of the cowling and the propeller hub were painted blue as on his other aircraft to represent his cavalry unit (Profile 20). At the same time he still had his old Albatros D III with a wine red fuselage. After several wing failures in the Albatros D Vs from this construction batch, they were looked upon with great scepticism by the pilots of the Staffel. When possible, the old Albatros D IIIs were flown in place of the new D Vs. Hans-Georg von der Marwitz also may have preferred to fly his patrols with his old wine red Albatros D III.

At first Otto Fuchs had had the idea of painting the fuselage of his Albatros D III black out of respect for his original artillery unit, but Hans-Georg von der Marwitz shook his head laughing: *"If you do that, then you'll get a challenge from Lt. von Bertrab."*[32]

Above: On the fence of the readiness hut. From left: Paul Erbguth, Frl. Kuhlenkamp, Hans-Georg von der Marwitz, Hans Bethge, Ilse von der Marwitz, Douglas Schnorr, Rudolf von der Horst, Kurt Katzenstein, and Karl Weltz.

This "threat" was of course not meant seriously, but as a result he had to choose a different color. Since yellow was taken by Lt. Heinrich Brügman and red was taken by von der Marwitz, he decided to paint the fuselage of his Albatros green. Lt. Paul Erbguth had decided to paint his Albatros D V in the Saxon colors of green and white, but Otto Fuchs made an agreement with the sociable Saxon that he would not use a bright grass green like he did, but rather a dark "bilious" green.[33] One of the authors reconstructed the "bilious green" color together with Otto Fuchs. While doing so, it was determined that the color "emerald green" approximately corresponded to the "bilious green" of the airplane (Profile 21).

Later he wrote about the appearance of his aircraft: *"Around noon a bright green bird is on display amidst the row of Albatros fighters standing ready for take-off. It looks like a big grasshopper."*[34]

As Otto Fuchs reported, in the summer of 1917 there were considerations for introducing a uniform letter marking, which however was never implemented, as most pilots preferred their individually painted aircraft, which were also better recognized in the air than the red letters. When in July Otto Fuchs took over Albatros D V D.2140/17, he applied a red "F" to both sides of the fuselage. But he had such little luck with this aircraft that he subsequently returned again to his green paint scheme (Profile 22).[35]

Albatros D III D.791/17 had a – supposedly – white cross stripe on the fuselage (Profile 23). Albatros D III D.2304/16, the old aircraft of Lt. Seitz, had the letter "S" over-painted in a dark oval (Profile 24). For the color draft "mauve" was used, but this is only an extrapolation. The reason to use this color was the statement of members of the Jagdstaffeln that colors used for the camouflage (e.g. mauve, dark green, pale green, rust red, and others) were available at the Armee-Flug-Park and used by the Staffeln for repair work and painting. Both aircraft had also the rudder over-painted. For the same reason we used camouflage green for the rudder. The identity of the two pilots is not known.

3. September 1917 – October 1917

Above: Pilots of the Staffel in October 1917. From left: Vzfw. Hans Oberländer, Uffz. Emil Liebert, Lt. Otto Fuchs, Lt. Kurt Katzenstein, Lt. Rudolf von der Horst, Lt. Douglas Schnorr, Lt. Karl Weltz, Oblt. Hans Bethge, Vzfw. Josef Heiligers, Oblt. Kurt Preissler, Vzfw. Josef Funk, Lt. Hans Holthusen, Lt. Paul Erbguth, Lt. Hans-Georg von der Marwitz.

An Eventful Autumn

By the beginning of September 1917 Jagdstaffel 30 was completely equipped with Albatros D Vs. Apart from that, there were still one or two Albatros D IIIs for newcomers to make practice flights.[1]

On September 4, 1917 the Jagdgruppe of the German 6th Army, consisting of Jagdstaffeln 12, 30, and 37, was formed under the command of Oblt. Bethge. The intention behind this measure was a concentration of force with the greatest possible flexibility. For that purpose, several Jagdstaffeln were combined under the leadership of the Staffel commander with the most seniority or with the greatest amount of operational experience. The Jagdgruppenführer (leader of the Jagdgruppe) had to coordinate the deployment of the Staffeln in such a way that more sizeable aerial incursions by the enemy would be confronted in the most unified fashion possible, thereby increasing one's own chances of success while simultaneously decreasing the risk of losses.

In the interim the military situation had made this necessary. Since the summer of 1917 the German Jagdstaffeln were not only numerically inferior to their Allied opponents, but the new Allied fighter planes and even Allied two-seaters, such as the Bristol Fighter, were superior to the German Albatros D III and D V. In this situation this was the only possible short-term step in the right direction, which bit by bit was employed on all sectors of the front.

By September the pilots of Jagdstaffel 30 had become a sort of blood brotherhood, both while in action and on the ground. Their conduct was more reminiscent of a high-spirited group of students than a military unit in which "order and discipline" reigned. The members of the Staffel regularly took "organizational trips" to Lille for the sake of good rations or to "organize" other necessary things, such as acquiring paint for the airplanes. Hans Bethge did not just indulge them, but also joined them. Douglas Schnorr also had the task of securing tickets for the theater, concerts, or music hall. Such pastimes were an important psychological means of reducing stress before and after missions.

In addition, Hans Bethge was prepared at all times to assist a pilot by talking to him if he were having difficulty coping with the stress of frontline duty or a particular situation involving aerial combat.

Above: Oblt. Hans Bethge in the pilot's seat of his Maus ("Mouse") before take-off.

Above: Otto Fuchs' insignia "Fox and Cock." The picture shows the insignia on his Albatros D III at Jasta 77. It was absolutely identical to that on the Albatros D V flown with Jagdstaffel 30 in the late summer/autumn of 1917. But at Jasta 30 the fuselage of the Albatros D V was "verdigris green".

Otto Fuchs had just such a conversation with him. He had taken off for a frontline patrol in splendid flying weather. While still on the German side of the lines he witnessed an impressive display of nature, a play of light with cloud and sun. He was so fascinated by it that he completely forgot the war for a few minutes and lost himself in the experience. He was dreaming to himself till suddenly he noticed that a British fighter plane was flying above him to the side whose pilot was just as distracted by the natural display as he.

Fuchs approached the enemy aircraft along the shortest distance and took a position beneath his tail without being noticed. Reflecting upon his own situation a few minutes previously, where he too would have been completely defenseless against an attack, he was suddenly subject to scruples about pressing the buttons of his two machine-guns. He turned away and allowed the "Englishman" to fly onwards unmolested, concerning which he reported:

At first I felt very good. I was proud of myself and regarded my behavior as ethically correct. The closer I came to the airfield, the greater became my doubt. What if in the future this Englishman were to shoot down a German airplane, would I then be responsible for the death of this comrade? Had I betrayed my comrades, did I not have enough courage, was I unsuited to be a fighter pilot?

This thought pursued me into my dreams. For several days I snuck around Hans Bethge until I gathered the courage to talk with him. I expected reproaches from him, but there weren't any. He sat down together with me in order to look at the problem from various angles. Finally, he asked me whether I would have opened fire if my maneuver had been observed by German soldiers on the ground or if a German two-seater had been in the vicinity.

I answered him with an unequivocal "yes." For me this question was a key to coming to terms with this experience. It dawned on me that there was no black and white, no absolute right or wrong, and what's more there is no universally valid morals in war.

Hans Bethge told me that he too had surprised

Left: The wine red Albatros D V of Lt. Hans-Georg von der Marwitz.

enemies and shot them down without them defending themselves. He probably didn't feel good about it, even if he had thereby satisfied his soldierly duty. Every time there was a difficult aerial combat with an opponent he felt relieved and regarded this aerial combat in which his opponent had every chance of shooting him down as well as a sort of balance, a sort of moral compensation vis-à-vis his conscience.

I became conscious that there was no definitive answer for this dilemma in wartime, but after the conversation I felt better. Shortly thereafter I had my second and third victories.[2]

On the evening of September 4, 1917 a formation of four D.H.4s of 25 Squadron RFC bombed Hantay. On the return trip they were taken on by five Albatros D Vs of Jagdstaffel 30 over La Bassée, who however identified the bombers as "Bristol Fighters." During this encounter Oblt. Bethge shot down a D.H.4 after a short combat over German territory near Auchy. In the meantime Lt. Kurt Katzenstein had attacked a second bomber and pursued it over enemy lines. Oblt. Bethge joined in, likewise attacking this machine, and caused it to crash west of Auchy.

Five days later Uffz. Emil Liebert arrived at the Staffel. He was an experienced pilot who had served with Flg.Abt.(A) 264 from January to August 1917. One of his missions nearly had a fatal ending for him. On April 30, 1917 he had taken off with observer Lt. August Rodenbeck, who had been detached to his unit from Flg.Abt.(A) 233, for an operational flight. Shortly after take-off they were attacked by several enemy fighters at an altitude of only 200 meters. In the process Lt. Rodenbeck was fatally wounded, while Emil Liebert managed with great difficulty to make an emergency landing in the badly damaged A.E.G. C.IV near Izel-lès-Ésquerchin.

It was probably for this accomplishment that he was awarded the Iron Cross 2nd Class, and after his promotion to Unteroffizier he reported to the fighter branch of the air service. Otto Fuchs, who in the fall of 1917 was Emil Liebert's "Ketten" commander, especially appreciated his courage and readiness for action. For this reason, he spoke favorably about him to Hans Bethge several times.

On September 17, 1917 Ketten from Jagdstaffeln 30 and 37 attacked a formation of D.H.5s from 41 Squadron RFC around 8:00 southwest of Douai. Uffz. Emil Liebert attacked a D.H.5 in a dive from superior height, which under his fire crashed a few minutes later north of Vitry-en-Artois. Shortly before Lt. Udet (Jagdstaffel 37) caused a second D.H.5 to crash.

On the next day there was again a fight between a

Kette from the Staffel and the same D.H.5 squadron. A "Sopwith single-seater" attacked by Oblt. Bethge exploded in the air and the wreckage hit the ground between the lines.

Four days later, on September 21, 1917, a Kette from Jagdstaffel 30 under the leadership of Lt. Paul Erbguth, together with a Kette from Jagdstaffel 12, attacked a formation of Sopwith single-seaters around 9:15. Lt. Katzenstein attacked, together with Gefr. Ulrich Neckel, Sopwith Pup A7321 and forced the machine to land. The victory was however granted to Ulrich Neckel from Jagdstaffel 12.

After taking off once more, Lt. Erbguth's "Kette" again attacked a formation of Sopwith single-seaters. After a long dogfight Erbguth succeeded in getting behind Camel B3914 of 45 Squadron RFC and forcing the airplane to land on a meadow near Wevelghem. Paul Erbguth landed his Albatros next to the "Englishman" who had made an emergency landing. When he climbed out of his machine, he noticed that his opponent clambered out of his plane unwounded. Paul Erbguth approached his opponent and offered him a cigarette. He saw to it that the British pilot, 2/Lt. E A Cooke, was brought to Phalempin, where he remained for a day as a guest of the Staffel. When the military police appeared to take 2/Lt. Cooke to the P.O.W. camp, he gave Paul Erbguth his flying gloves with the words "good luck." The latter did use them through the entire war. After the war, he took the gloves home with him as a souvenir, which had in fact brought him good luck.[3]

On one of the following days Paul Erbguth and Hans-Georg von der Marwitz took off in the direction of the front. They crossed the lines and over the nearest Allied airfield dropped the message with a small parachute that Lt. Cooke had fallen into German captivity uninjured. In a conversation with Paul Erbguth, he had expressed concern that his parents might think he was dead. Paul Erbguth wanted to spare the parents of the nice "Englishman" this sorrow.

On October 2nd there was an aerial combat between a Kette from the Staffel and some Bristol Fighters of 11 Squadron RFC over Marquette. In the course of the fight Vzfw. Hans Oberländer was able to shoot down a British aircraft at 18:10. Upon impact the British machine was completely wrecked.

Four days later Gefr. Josef Funk spotted a Spad around 10:30 which had become separated from its formation, having apparently gotten lost, and was flying alone at 500 meters altitude. Josef Funk attacked the enemy, who thereafter landed southwest of Seclin. The British pilot had been wounded during the attack and fell into German captivity.

Above: Lt. Rudolf von der Horst zu Hollwinkel with his fiancée on leave in Berlin in the autumn of 1917.

On the evening of the same day Lt. Hans Holthusen arrived at the Staffel. He was born on April 13, 1894 as the son of a merchant in Hamburg. After his Abitur he followed his father's wishes regarding his profession and became a merchant like him in his firm. At war's outbreak he went to the front as a volunteer with Reserve Infantry Regiment 90.

After his switch to the air service he was transferred on June 26, 1916 to FEA 4 in order to be trained as a pilot there. Afterwards he proceeded to Flg.Abt. 18, which was under the direct command of A.O.K. 6. With this unit he completed a whole series of dangerous long-range reconnaissance flights. However, his flying career nearly came to an end on August 21, 1917, as witnessed by a report from Flg. Abt. 18:

At 5:50 a.m. the crew of Lt. Holthusen and Lt. Schröder took off in Ru. [Rumpler] C IV 8439/16 for a long-range reconnaissance. Fired upon on the return flight by our own flak. Between Ascq and Thumeries attacked at 700 m altitude by Vzfw. Fahlke, Jagdstaffel 37, in Albatros D III 2127/16, which took off due to the flak fire. During this pilot Lt. Holthusen was hit by a ricochet in the left temple and Lt. Schröder was hit by a round through the right side of his chest and in the left hand.[4]

This experience may have been the reason why

Above: Lt. Rudolf von der Horst climbs out of his "sugar egg" after a mission.

Holthusen transferred to the fighter arm. His observer, Lt. Schröder, recovered from his wound and returned to Flg. Abt. 18. On February 26, 1918 he visited Jagdstaffel 30 as the guest of his former pilot.

On September 4, 1917 Hans Holthusen was ordered to the Jagdstaffelschule at Valenciennes. In his flight log the following flights are listed, which provide a good insight into the four-day training progam of the school at this time:[5]

Date	Aircraft	Flight Purpose
29.09.17	Halb. D II D.160/16	Practice flight
01.10.17	Alb. D II D.506/16	Five practice flights
01.10.17	Alb. D I D.423/16	Practice formation flight
01.10.17	Alb. D I D.423/16	Aerial combat practice, two flights
02.10.17	Alb. D I D.423/16	Turning practice flight
02.10.17	Alb. D II D.1078/16	Practice formation flight
02.10.17	Alb. D III D.2122/16	Aerial combat practice
02.10.17	Alb. D I D.423/16	Turning practice flight
02.10.17	Alb. D III D.2028/16	Target practice flight
03.10.17	Alb. D III D.2028/16	Turning practice flight
03.10.17	Alb. D II D.1773/16	Target practice flight
03.10.17	Alb. D III D.1925/16	Firing practice flight

Two days after his arrival, on October 8, 1917, Hans Holthusen completed his first 18-minute familiarization flight with a maximum altitude of 2400 m. over the Staffel's airfield in an old Albatros D III D.799/17. The day after followed some practice circuits with machine-gun fire in the Albatros D V D.2199/17 assigned to him.

A little later he took off in this machine for his first operational flight, which however he had to break off 8 minutes later due to technical difficulties. After the repair and a further test flight he took off on October 10 with this machine for an operational flight with a duration of 35 minutes, during which he was fired upon by British flak.

Otto Fuchs described Hans Holthusen as a friendly man from the Hanseatic region, brave, correct, reflective, extremely reliable, and absolutely trustworthy. Besides that, he was a good singer who cheered up his comrades on many evenings with his songs. Equally appreciated were his talents as an amateur kitchen chef. He was also considered the gourmet of the Staffel.

On October 11, 1917 Hans Holthusen was one of the witnesses of an unusual event at the Staffel's airfield. Two Ketten of Jagdstaffel 30 had returned from a patrol and the pilots sat in the little readiness hut on stand-by. Two further Ketten, one led by Otto Fuchs, were still in the air. Suddenly they heard the sounds of an engine of an approaching aircraft.

The pilots went out on the terrace and to their

great astonishment saw a Sopwith Camel, which slowly glided towards the field, landed, and taxied. The British cockades could be seen clearly on the airplane. The Germans ran as fast as they could to the machine which had just landed. The pilot of the British airplane climbed out and walked towards the men coming up to him. They had their flying outfits on and therefore could not be recognized as Germans at first sight.

The British pilot – it was 2/Lt. W H Winter of 28 Squadron RFC – politely asked for "essence" [Fr. "fuel"] and "castrol oil" in order to be able to continue flying. When he received a reply in German to his question, he was completely taken aback. Lt. Douglas Schnorr, who because of his wooden leg was the last to arrive, explained to him then in his best English that he was on a German airfield and that he was now his prisoner.

When Otto Fuchs landed a little later with his Kette he was more than surprised to find a flightworthy Sopwith Camel on the field. On the next day the Camel's cockades were overpainted with the German national insignia and the machine fueled up, so that the pilots could test the machine. These flights had to be reported to the Kofl and the Camel was always accompanied by an Albatros from the Staffel for protection. This was to prevent the "German" Sopwith from being accidentally attacked by German airplanes. These flights gave the pilots valuable experience regarding the flying characteristics of the Sopwith Camel.

As for the rest, 2/Lt. Winter, as previously with some of his other "colleagues from the other field post address," was able to enjoy the hospitality of the Staffel as compensation for his mishap before he was turned over to the military police.

Between October 10th and 15th Lt. Holthusen undertook eight patrols during which there was no contact with the enemy. On October 16, 1917 Hans Holthusen received "his" own personal machine, Albatros D V D.2126/17, in which he immediately undertook a flight behind the front in order to give it a thorough "shake down". Two days later he recorded in his flight log his first contact with the enemy while with Jagdstaffel 30:

Alb. D V 2126/17, morning, overcast, flight duration: 65 minutes, maximum altitude 4,800 meters, flight path 4th Army. An aerial combat with 9 Sopwiths on the other side of the lines near Zonnebeke, fired upon by flak.[6]

On the morning of October 21, 1917 Lt. Holthusen was once more operating with some Albatros from the Staffel in the zone of the 4th Army. Near Armentières there was an aerial combat with four aircraft designated as "Sopwiths". In the afternoon of the same day operating in the zone of the German 6th Army, the Kette along with Hans Holthusen attacked two R.E.8s, which were apparently protected by four "Sopwiths". Both missions were without losses for either side.

Above: Lt. Kurt Katzenstein in the garden of the Staffel quarters. (Stiftung Deutsches Technikmuseum Berlin/ Katzenstein Collection)

On the other hand, the following day Uffz. Liebert succeeded in forcing a Bristol Fighter to land, which had previously been fired upon by Flakzug 27. The unwounded crew set their airplane on fire after the emergency landing. However, the victory was not granted to Emil Liebert, but rather to the anti-aircraft battery.

On October 27, 1917 Otto Fuchs took off with

Above: Lt. Kurt Katzenstein in his Albatros D V with the "cat on a stone." (Stiftung Deutsches Technikmuseum Berlin/ Katzenstein Collection)

Rudolf von der Horst and one other pilot at 10:00, about which he reported in his book *Wir Flieger*:

I was flying along in the late afternoon [in the morning—authors] close behind the gray blanket of rain clouds, to the right of me an airplane from the Staffel, to the left von der Horst. Suddenly, near Lens, three Englishmen storm out of the clouds right towards us. One of them buzzes so closely above me that I feel the pressure of the air squeezed between his and my wings on my eardrum.

They are Sopwith single-seaters led by an aircraft bearing streamers. Three against three. An even match. Good! On our side of the lines ... All the better. I suddenly tear my machine around and am sitting on the neck of the last one. Far below I notice the one airplane from my Staffel ... He is clearing off. Von der Horst is following him in sharp searching turns. He obviously does not know what is going on. I am alone...

At two or three aircraft lengths I aim at the light brown leather cap in the enemy machine, calmly, without trembling. Clack-clack ... Clack-clack-clack ... The puppet collapses in his seat, the machine tilts forward and, turning its wheels and the light undersides of its wings upwards, shoots past below me. Is it a feint? I think of the Nieuport recently ... Rearing up, with a half-roll I stand on my nose, once more behind my opponent. Not necessary. Swirling like a maple seed, he bores into the depths trailing a long, white trail of smoke.

The next ... Where? Two cockades disappear above me in the lead-gray densely clustered mist ... Above the front two black rounds of flak explode. The silhouette of the third Sopwith is making for a broad green streak of light breaking through the

Above: Vzfw. Josef Heiligers lands in his Albatros D V "red X."

clouds there. Is that all? Yes, that is all. I look below me where my victim is just crashing. There is a cloud of dust. Then, after it has dispersed I see a little heap of rubbish. I descend in large spirals.

A small, narrow meadow runs a hundred paces next to the site of the wreckage. After three futile approaches I land. I go to where groups of curious onlookers are gathered. The circle parts before me, lets me in, and closes like a maw.

There he lies ... A formless lump. A ground-sheet is covering his head. Pointed bones poke through his bloody leather jacket. A broad blackish red puddle runs over crumpled stretches of linen and is sprayed over the confused heaps of splintered framework, glaring cockades, and steel wires. The engine is sticking a meter into the ground. It smells like castor oil, squashed grass, and warm blood. I stare absent-mindedly at all of this. My handiwork! Hmm ... !

Someone seizes my hand impetuously. What? Did I hear right?

"Congratulations, comrade ... A spirited attack..."

And he presses the wallet someone had taken off the dead man into my hand, tears it away from me again since my stiff fingers are holding it with indifference, and begins to unfold it excitedly before my eyes, during which he states:

"... A picture of him, take a look ... Strapping fellow, by golly ... Letters, rose petals, small image of a saint ... Aha! Look, look ... Nice girl, eh? It's too bad for the poor thing ... Pay book: 6th May 'ninety-five, so twenty-three years old ... What else does he have in here? Souvenirs of Paris ... Oh my ... And there ... Bank notes, three, four pounds ... that's all? Thin ... thin ... There ... So . . ." He stuffs the whole lot back into the wallet and hands it to me again.

Disgusted by the excited demeanor of the strange person, I refuse, turn towards the wreckage, pull out my pocket knife, and cut out a cockade. The other man is still talking insistently to me.

"But you have to submit a report! Don't you understand? So please, here!"

As I roll up the piece of linen and trot towards my airplane, he follows after me until I take the leather wallet, place it on the seat, and then sit on it.

My motor starts up. I rush hopping across a piece of the meadow, rise from the earth, go into a steep turn above the scene of the misfortune, and race at the height of the treetops to the airfield.[7]

The three aircraft described by Otto Fuchs as

Above: Fun on the airfield. From left: Hans-Georg von der Marwitz, Kurt Katzenstein, and Hans Holthusen in October 1917.

"Sopwith single-seaters" were a flight of Sopwith Camels from 43 Squadron RFC consisting of Capt. A G Fellowes, 2/Lt. R Harris, and 2/Lt. G P Bradley. The machine attacked by Otto Fuchs was the Camel B6374 of 2/Lt. Bradley.

In the official War Diary of the Royal Flying Corps it is tersely reported about this air combat (in parenthesis are additions from the pilot's combat report):

Capt. A G Fellowes, 43 Squadron, attacked a D.F.W. two-seater (at 9:35 English time in Camel B2367) which he brought down out of control (near Sainghin). Afterwards he attacked an Albatros single-seater (at 09:45 English time), fired some bursts at him and therewith caused him to crash in flames (near Sallaumines).[8]

On the other hand, in the squadron history by J. Beedle, it sounds much more warlike:

... Captain Fellowes, who had already destroyed a D.F.W. reconnaissance two-seater, fell like an avenging angel on a Fokker [!] scout which was following down its victim [the Camel flown by 2/Lt. Bradley—authors] and shot him down in flames.[9]

The German two-seater concerned was a machine of Flg.Abt.(A) 235 with the crew of pilot Uffz. Graessner and observer Lt. Koch, who made an emergency crash landing in the vicinity of Harnes. The pilot was badly injured and the observer suffered minor injuries. The Albatros reported as "shot down in flames," which in a third source is only described as "crashed," was in fact the downed Camel. This is a clear example of how easily one can be confused "in the heat of battle."

After landing next to the downed Camel, Otto Fuchs took off with no problems from the meadow and landed smoothly at Phalempin. There Otto was cheered by his comrades and mechanics, but because of the circumstances of the aerial combat was in no mood for celebrating. He simply wanted to be alone and withdrew to his room, where he lay down on his bed, and after some time was disturbed by Hans Bethge.

...Then there is a knock.

"Yes, what is it?"

Bethge comes in. I jump up.

"Relax, for heaven's sake! Remain resting on your laurels!"

The large glasses peer merrily and with so much

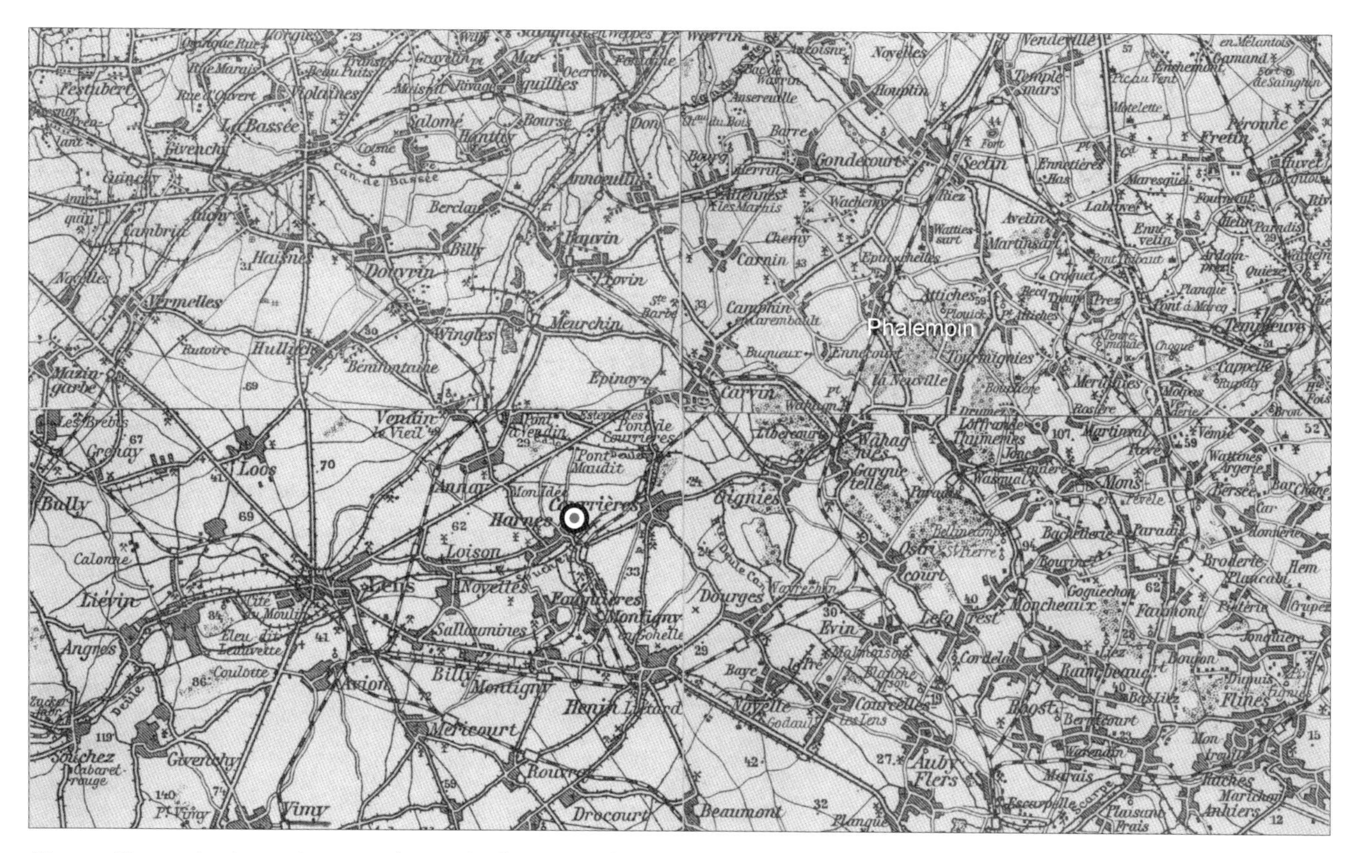

Above: The cockade on the map shows the location of Lt. Otto Fuchs' victory on October 27, 1917.

warmth that my annoyance evaporates without a trace.

"I have to pay a visit to the little hero of the air and congratulate him, as is fitting. By the way, I've also brought something for you."

He unfastens his Iron Cross, First Class from his tunic and pins it on my breast as I stand speechless and my head fills with fire.

"So . . ." he says and takes a couple steps back to observe his work with satisfaction. "A beautiful medal ... But now you've put on a different face! There, look in the mirror, eh? Sickening, isn't it, Madame? Is that the countenance of a fighter pilot? Hand me the comb, hair oil ... none there? What's that there ... Nothing ... And this ... Ugh!"

He fumbles around on my washstand, rummages through the drawers, sniffs the empty little bottles which belonged to my predecessor, knocks over my water carafe, "she also has a toothbrush, the sow..." until I shove him away laughing and push him into the armchair.

"Now just sit here until I'm ready and not another peep out of you!"

"Ha, what do you think you're doing?"

"Not another word, okay?"

"This is how he treats his Staffel commander in person!"

"He deserves no better. If you'd gone through what I did today . . ."

"How did you fare then, you ridiculous raven? In the name of God, spit it out!"

I tell him the entire course of the encounter with the three Englishmen and how in an inglorious way I achieved my success, that I simply knocked down this sleepyhead at close range, ambushed so to speak, mean, raw, beastly ... that since then everything annoyed me, from the officer who palmed the wallet off on me to the Iron Cross, First Class.

"Hmm," Bethge says and peers for a while towards the window. "First of all," he finally begins, "I would like to hear how the other pilot and von der Horst conducted themselves during the affair."

"The other pilot (Fuchs never told his name – the authors) and . .. yeah, right ... His motor oiled up."

"Curious. Really curious ... No complaints against von der Horst?"

I shake my head: "No".

"And now, regarding your situation: you stood thus alone against three."

"Notwithstanding."

"Well, get a load of this arrogance. Admittedly I cannot help you with that."

"You could really ruin one's sense of humor

Above: On October 6, 1917 Spad VII B3508 "C" of 19 Squadron RFC was forced to land by Gefr. Josef Funk.

if they allowed it," he says, picking up the conversation once more. "Of my seventeen victories—I can confide this to you—at least ten of them took place in a manner similar to yours today. I believe generally that's the way it is with most fighter pilots. Only they're sensible enough not to grow any gray hairs over it. To dismiss cunning in wartime means to shut off one's brain and make success the privilege of stupidity. So one surprises one's opponent and gives him the works before he can say 'boo.' There is rarely an honorable combat in which only daring, skillful flying, superior aiming, and good nerves determine the outcome. For the most part chance is involved and can cost the most capable flyer his life.

"There are a thousand vulnerable spots on one's crate and two-seaters in particular are a nasty invention. They can defend themselves on all sides like a porcupine, while such as we can only bang away towards the front. You yourself have been sent home often enough all shot-up. That's why I say the normal approach for us is one of surprise.

"Resolve, a dive, approach up to fifty meters, back sight, fore sight, the Englishman's skull, eyes closed, and press the button ... Bang, it's already blown apart. At first I was also occupied with the moral side of the matter. What you're doing is actually reprehensible, I thought ... But then I thought again: They've shot you up time and again. Why shouldn't you be able to do the same thing? In that way the moral balance is to some extent reestablished. Do you understand what I'm saying?"

"Yes, but . . ."

"'Thought!' I would ask the gentleman to consider that I said 'thought.' Amongst other thoughts this one occurred to me as well. I admit it's very convenient, superficial, and for a more demanding conscience absolutely inadequate. What can you do? At times I harbor such primitive and inadequate thoughts."

He suddenly gets up.

"You're mistaken, my dear fellow, if you think I'm going to prattle away at you for hours on end. You novice, you tender little lamb! Try shooting down a dozen before developing scruples about whether or not it's the right thing to do, got it? March on down to the mess now . . ." He pushes me out the door with a few jabs. 'And put on the face of a fighter pilot, if I may so request. We have guests.'

"You know, you're a real monster, Bethge..."[10]

Otto went with Hans Bethge into the mess, where he was met with a great hullabaloo. Because some guests had been invited that day Hans Holthusen, who functioned as the kitchen chef, had gone shopping in Lille and prepared a large meal. In the end the evening of merriment helped Otto Fuchs get over his gloomy thoughts.

As Otto Fuchs later declared during a conversation, he was by no means the only fighter pilot who was concerned about the moral correctness of his actions. However, it varied from

Above: Oblt. Bethge and his Offizier z.b.V. adding to the Staffel's trophy collection on the roof of the hangar. The trophy collection led Lt. Winter to mistakenly assume that it was an Allied airfield.

person to person. Many coped with the moral burden only with difficulty while others were set on their personal ambition. They had what one called in flying circles "Halsschmerzen" [a sore throat], which meant that they were absolutely determined to acquire the "Halsorden" Pour le Mérite as an allusion to the medal being worn at one's throat] and simply pushed aside moral doubts. Again others, like Hans-Georg von der Marwitz, simply did not appear to take life so hard, for which reason Otto sometimes silently envied him.

With regards to the conversation, Hans Bethge's statement that two-seaters were as a rule more difficult opponents than single-seater fighters is interesting. This observation was confirmed for the authors by other German fighter pilots. The Bristol Fighter in particular was considered as an extremely dangerous opponent. At this point it should once more be noted that the primary task of the fighter pilot was to do combat with enemy two-seaters (reconnaissance, artillery observation, and bomber aircraft) and to hinder them in carrying out their mission.

Two days later, on October 29, Otto Fuchs was waiting together with other pilots from the Staffel in the readiness hut for the report of enemy aircraft from the front. It had rained the entire night and in the early morning heavy rain showers had alternated with hail so that at first they were condemned to inactivity. Around 10:00 came the report of lively enemy aerial activity by the Souchez and Vimy Groups located to the southwest of Phalempin. The Staffel took off first in a northerly direction towards Seclin, in order to then fly in a westerly direction towards the front. There the Albatros aircraft turned towards the south in the direction of Souchez. In the vicinity of the city of Lens Oblt. Bethge spotted aircraft over the front, concerning which Otto Fuchs reported:

Ha! Five little black dots, no bigger than the heads of pins. Bethge, you are one awfully clever animal to see something like that! And of course he had seen it the whole time.

Where are we then? I have to lean far overboard and look rearwards beneath the tail in order to find the earth at all. In front of me and beneath me a single shimmering mudflat spreads out. But there behind it is Lille ... flat and inconspicuous like an old copper coin. Going by that we are floating approximately over Fleurbaix, perhaps even a little to the west of it, in any event above the front.

Above: Sopwith Camel B6314 of 28 Squadron RFC shortly after the mistaken landing on the Staffel's airfield.

What are the five dots doing? I am looking... looking... there! They have become even tinier. They are flying southwards. In the vicinity of Béthune... Whoa! Watch out!!! I yank my machine into a sudden left turn over von der Marwitz' wine red Albatros. He did not notice Bethge's turning movement. That had to happen. Of course. Now the five are floating obliquely in front of us and considerably lower. We have clambered up to 4300 meters and now make use of our height and engine power to attain greater speed. What we have in the way of superior speed is applied in a westerly direction. So the distance between them and us remains the same, but we are slowly shifting farther in the direction of enemy territory and therewith gain a double advantage. On the one hand the enemy is left in a state of uncertainty as to our nationality. But on the other hand if he does recognize us as Germans it will give him an impression of timidity so that he for his part will feel a greater desire for combat. For with such a west wind one must not offend the Tommy with a crude attack, otherwise he will immediately dive into the clouds... And we so very much like having him above them!

For certain he is assessing our refined discretion. How could he act differently now and fly towards the east? Of course this is just a ruse on his part. A crafty rogue, the enemy squadron commander! A few minutes later he has already turned around again. He does not trust the peace. Nor the fifteen kilometers distance. He knows that if we were to rush towards one another it would take hardly two minutes to cover this distance. But Bethge also knows this and so he acts as though the five dots do not concern him at all.

The other leader repeats his experiment and turns towards Germany. He appears to have serious doubts about our origins. To be sure he must be thinking: if those seven over there are Germans, then they are cowardly if they do not attack me now. But if they are Englishmen, well, then I can feel free to get a bit cheeky! However, in an emergency I do have the clouds available... And he flies further eastwards than he can actually justify.

He turns again. Rrr-eee ... We too are lying on our wing tips. Now please, no hunting fever! Are the guns in order? My thumb is pressing on the button on the control column. Tack-tack says the right one. Tack-tack-tack says the left one with flawless pronunciation. Good! Also a searching glance at the altimeter: four thousand three hundred. At the rev counter: fifteen hundred. Fuel pressure: two. Over the entire machine ... and then rearwards up in the sky. As the last in the formation I am responsible for surprises from that direction. The air is clear all around.

And now to our Tommies! The five dots rapidly

Right: Hans Bethge inspects the trophy collection on the roof of the hangar. Behind him the captured rudder of F. E. 8 A4887 of 41 Squadron RFC can be seen. (Stiftung Deutsches Technikmuseum Berlin/Historisches Archiv)

change into five flies. The five flies suddenly become five butterflies with bright peacock eyes on their wings: Sopwith single-seaters just like the day before yesterday. Just no hunting fever! One hits nothing if one takes it too seriously. Why does Bethge not attack? I hop around on the leather cushion with impatience, straighten out my goggles, and squeeze my left eye shut for aiming. In front of me von der Marwitz wags his rudder and Holthusen's single-seater hangs noticeably towards the side on which his leader is swerving.

The Englishmen are just passing a thousand meters below us. Bethge throws his machine onto its wing. The other six of us do the same. The Englishmen do not move. We gather closely behind our leader. My nerves are tingling. Everyone is waiting for the red flare. One can hardly stand it.

Finally! A long white thread of smoke with a red dot at its tip separates itself from Bethge's Albatros. I tilt forwards at full throttle.

I pull my head in, hold the stick calmly, and eagerly listen as the howling of the wires gets higher and higher. The one hundred and sixty horses are stomping and shaking in their narrow housing, a roaring whirl of air whips against my forehead and tears at my cap. Hans-Georg von der Marwitz stays back behind me... Then Holthusen. Slowly I overtake one after the other. I push past another pilot... the enemy machines are growing. Now I am flying close by Bethge.

Then the Englishmen also tilt. In an instant all five of them are standing on their noses and stretch their tails towards us. I push the control column forwards. A jolt goes through the fuselage. The cables screech yet more shrilly. My steel heart rages more wildly, more bitingly the cold current of air hisses around my ears. With a jump the rev counter climbs above sixteen hundred and gives out. Faster, faster! Now not even Bethge can keep up. He waves at me. I return the greeting. The hurricane nearly dislocates my arm. Three hundred meters still separate us from our opponents. Now and then they emit little oily clouds. It stinks like castor oil. They too are racing. Nearer and nearer draw the clouds in which salvation beckons.

The first has already disappeared therein, the second, the third, the fourth, ... the last... no, not the last one! Instead of following his comrades, he suddenly pulls his aircraft up, applies throttle, and attempts to swing behind me in an elegant arc. My turn comes out sharper. He notices it and changes his turn. Idiot! For an instant he hangs in the air in front of me like a target. I glance vertically from above into the body of the aircraft. Rear sight and

Above: Ilse von der Marwitz with several nurses while visiting her younger brother in October 1917. Standing from left: Lt. Hans-Georg von der Marwitz, Lt. Paul Erbguth, Lt. Hans Holthusen, "Nurse" Otto Fuchs, Lt. Rudolf von der Horst, Lt. Douglas Schnorr, Oblt. Kurt Grasshoff (commander of Jasta 37), Hans Bethge. Sitting, from left: Frl. von Arnim, Frl. von der Marwitz, Frl. Kuhlenkamp, Frau Kuhlenkamp. (Erbguth family)

fore sight, a light red leather cap... He throws his arms up.

I spring over him with a bound, and throw myself around again. The Sopwith is diving, slowly turning over. Whitish fuel vapor streams from him, then turns dark. The rest is concealed by the cloud.

One minute after that my comrades swarm around me. They wave to me, turn wildly around me, fire flare cartridges into the air, and demonstrate their approval in every conceivable way. I place myself next to Bethge, who is now heading for home. The others array themselves behind him.

While levelling out a gust pushes me to the ground too soon... Rumm... I bounce up again, have to quickly apply the throttle and fly another circuit. In the meantime Bethge waddles towards his hangar like a duck.

Von der Marwitz sets down in front of me. His left wing brushes the ground, he goes into a sharp curve, and then stands on his nose. Mechanics come running.

Again my bird touches the grassy surface and rolls rumbling to a standstill. Guided by a helpful crew, I buzz the engine until close to the stall, where Bethge is waiting for me with a grin. The infantry had already reported the downing of the aircraft:

"English aircraft fell out of clouds in flames near Gavrelle after aerial combat. The wreckage lies behind the first enemy trench line."[11]

On the same day Otto Fuchs took off to visit his brother Rudolf at Flieger-Abteilung (A) 288 at Dorignies near Douai.

"After lunch I sit in my snarling Albatros and keep an eager look-out over the shaking cockpit wall for the rail line from Lille to Douai. It is raining in torrents. I flit closely above telegraph poles and tree tops. I want to visit Rudolf one more time before I am transferred. He said on the telephone beforehand that his training to become a fighter

Above: The Pfalz D III 4078/17 delivered to the Staffel is moved to the airfield to be assembled. (R. Absmeier)

pilot was approved. He has so very much wished to become a pilot. I am glad for him and sing during the entire flight to my goal, which I reach after twenty minutes."[12]

The successful missions like those illustrated above aside, most of the missions consisted of flights which were carried out without result, as Otto Fuchs describes:

Wrapped in my fur coat, my hands buried in the pockets, I carefully climb through the wet grass of the airfield, now and then pausing, yawning and blinking at the colorless morning sky with sleepy eyes. Early take-off. Four o'clock. There is nothing going on. Two more hours of sleep would have done me some good. If only the sun would at least appear soon and warm me up a bit. It is noticeably brisk and my stomach is growling.[13]

In the readiness hut Otto Fuchs meets up with two comrades, one of whom has fallen asleep in an armchair. He pages through the well-read newspapers and illustrated magazines lying around in the readiness hut. Thus seven o'clock arrives and the sun is coming up. Then the voice of the Unteroffizier rings out: "A Holy Ghost!" This was the name for the phosphorus bombs which English aircraft dropped on German captive balloons from a great height in order to send them up in flames. The observers hardly had a chance of surviving in the rain of phosphorus, even if they succeeded in jumping out with a parachute. They choked in the cloud of gas.

The mechanics, who have drifted off to sleep bent over in our Albatros fighters, waken with a start.

"Crank 'em up!"

The siren howls... The engines have started up. Pulling my coat around me, I climb into the cockpit and raise my hand as a signal for take-off. We roll beside each other with increasing speed across the expanse wet with dew. From wing and propeller the moist air flows like a thin liquid. At a steep angle we thrust into higher regions opalescent like mother-of-pearl. Light veils swirl past. Below on the earth rivers of milk and valleys of creeping vapor form. More deep and pure the eternal blue arches above us.

We take a direct heading towards the streamers of smoke which have unfurled above the zone of the German balloons. Fatigue and vexation have vanished. Eye and ear are focused on the noisy, snarling machine. All one's senses are alert. After hardly ten minutes the first tube-like remnants of the incendiary bombs pass by and we keep a sharp look-out for the enemy evildoers. After long efforts I do spot four pale dots which are disappearing westwards towards us from down below—entirely haphazardly of course and without danger, as no

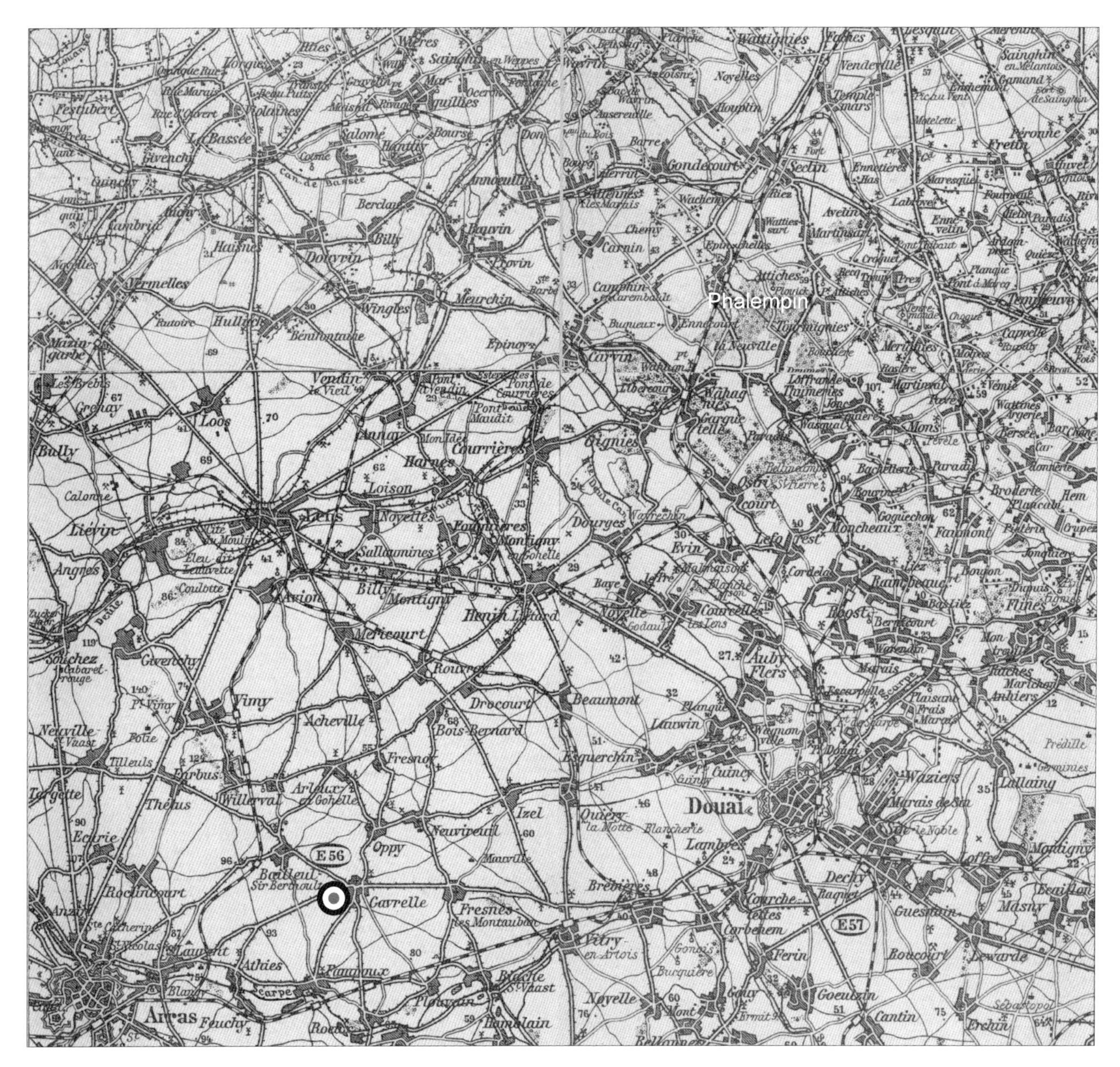

Above: The cockade on the map shows the location of Lt. Otto Fuchs' victory on October 29, 1917.

one can see us from the mist-covered ground.

After we have wandered around aimlessly over Armentières for half an hour I decide to turn back. What is the point of using up precious fuel and freezing one's bones? In spite of the advanced season I am quacking like a duck at the four thousand meters height we have climbed to in the meantime. So throttle back and head down! I fire off a green star-shell, which means: "We're flying home!"[14]

On the morning of October 31, 1917 Hans Holthusen, along with his flight, had first an aerial combat with four "Sopwiths" in the airspace of the German 6th Army and afterwards a fight with three aircraft described as "B.F.s" (Bristol Fighters) near La Bassée.

Around 12:30 Hans Bethge and Otto Fuchs took off together with four aircraft of the Staffel from Phalempin airfield in the direction of the front. Over Provin the Albatros D Vs turned towards the south, then the airplanes divided up. Otto Fuchs took a course towards Billy-Montigny:

"Because we have been up for just a short time I decide to add on a short stroll along the front. Beyond the range of the flak I begin to climb once

Above: One of the Pfalz D IIIs (4078/17) delivered to the Staffel between 21 and 27 October 1917 for testing at the front.

more. Karl Weltz accompanies me. Hans Bethge also joins up with us and indicates via flare signals that he himself is taking over the leadership. We fly southwards along the front, climbing steadfastly, so that I am more lying in my leather seat than sitting in it. My legs are pointing forwards, upwards, where the shiny nose of the machine is jutting high into the gentian blue sky.

When I turn around and look backwards over the slim verdigris green fuselage beyond the rounded tail, which seems to be firmly planted in the bluish and blondish depths, I am gripped by a slightly wondrous, sweetly eerie feeling. It is hard to believe that one can fly at such a steep angle. But the rev counter removes all doubt. And again my eye glides over the slightly trembling elevator towards the sunny abyss.

This view from the aircraft is the only thing which can give one a distant feeling of height. I suspect the reason for this lies in the slender lines of the fuselage which, though they cannot establish a connection to the earth, at least suggest one. If on the other hand I lean to the right and the left overboard, then the empty nothingness begins without transition and one's eye has no chance of forming a connection to the merry, colorful map below.

Incidentally, I love this cool boring and gnawing at the heart until suddenly the full recognition of the danger sets in, almost like the prick of a dagger. Slowly and uncannily it causes the soul, which weighs horror and delight on a precise scale, to grow pale. They are moments of extreme tension and even deathlike stillness, unforgettable eternal moments in which one's entire life crystallizes in a tiny point, in a luminous nothingness. They occur in almost every air battle, but also while trying out some sort of new aerobatic figures which the machine might not withstand.

I am convinced that these moments constitute the secret love of most flyers, at least the most daring of them. And even those who do not want to admit it, or really do not know what puzzling paths their tendencies tread, are proud of the mysterious exhalation of beauty and danger which surrounds their profession.

Interrupting myself in my reveries, I observe the patch of smoke near Gavrelle. It has gotten bigger. It looks like a fuming, boiling sea of lava. It is also growing upwards. Brownish veils of dust and nicotine yellow vapors reach till just below the clouds, which are sailing over individually like shimmering white swans, casting blue shadows on the soft blonde face.

Sooty spots ... six, seven ... and more over there, freshly flashing . .. on the German side. German flak! I search and search. Hardly visible through his earth brown camouflage and as big as an ant is an enemy aircraft. I estimate it to be fifteen hundred meters up. It is creeping eastwards, untroubled by the anti-aircraft fire.

Suddenly cut back, the roar of my engine turns into incomprehensible blubbering. The meter-long flames of unused gas shoot out of the exhaust, which bangs like a small mortar. The earth, like a large cake with the Englishmen as a middle-point moves into my direction of flight. I dive down almost vertically. In the thin layers of air above

Above: Lt. Holthusen with his dog "Puck."

five thousand the dive proceeds with unbelievable speed. By the time the indicator on the altimeter sinks to four thousand one can just take two deep breaths with an open mouth. Throwing my head all the way back to the nape of the neck, I see Bethge and Weltz flying straight ahead unswervingly. They still have not recognized what it is all about. In a minute I will show them.

I crouch quite small and compact behind my motor, blinking only now and again at my opponent, who is growing and growing. The wires are whistling, the wings tremble and shake. It is as though mountains were lying on my control surfaces, so mightily is the air pressing down on them. I rush past the enemy five hundred meters behind him. It is an R.E. (a British Reconnaissance-Experimental), a giant angular crate with an upper wing that juts far out.

I pull out and approach from below on the most favorable side. A thousand eyes are following me from the trenches, from the artillery positions. A thousand pounding hearts await the beginning of the drama ... Still two hundred meters distance. My thumbs are twitching, my teeth are boring into my lips. I plan to fire my burst into his body at very close range. He must fall! I keep him exactly in my line of sight and perform aiming exercises: a finger's breadth behind the motor, in the forward third of the fuselage. I get the shivers. Without the slightest movement, the proud R.E. is sailing straight ahead in front of me... My fourth!

At this moment Bethge falls from the sky like a stone, firing vertically into the Englishman from above. He wakes up and goes into a turn. The observer, taking me for the attacker, sends me a rattling greeting from his double-yoked machine-guns. He immediately recognizes his mistake. Bethge had sped past him to the side and even as he tackles him again, pouncing from below like a predatory cat, the observer spits a well-aimed series of bullets into his ribs. Bethge nimbly slips below him to the other side. With a jerk the "Franz" throws his pivoting machine-gun mount around. Little reddish tongues flicker before the mouths of his guns and a fine mist disperses. Bethge's guns speak as well. As can be seen by the desperate turns of the enemy machine, his speech is very distinct.

Then the Albatros tumbles... apparently in the slipstream... or has he been hit? No, not hit. After two uncontrolled somersaults he pulls out and even before I can intervene he is sitting on the other's neck. The enemy pilot immediately used the brief pause in the action to head for his lines. He senses that it is getting serious.

The heavy R.E. displays an unusual temperament. He pulls up his nose and with his wheels pointing upwards comes at me firing, though I am circling harmlessly fifty meters above the combatants. However, Bethge follows him into the loop at a distance of ten room lengths. All attempts of the Englishman to shake off the troublesome pursuer fail. Even when he follows a totally impossible flight path, beginning to turn like a spindle in horizontal flight and rolling away over his left wing, it does him no good. Bethge, who matches his aerial stunts, follows close on his heels.

I do not know what I should admire more, the tenacity of the defender who does not give up or the blind doggedness of the attacker who does not let his victim out of his clutches as long as he still has a round in his gun and a drop of blood in his heart. One of the two must fall. And the marrow freezes in my bones when I see the cold-blooded observer firing in the midst of these breakneck acrobatics, during which his head is often hanging downwards. The struggle has lasted over ten minutes without anything decisive occurring. Both are conscious of their ability and wrest the last measure of performance from their machines. They do so while hardly losing any height and without the earth below us shifting.

The efforts of the Englishman to move the battle towards the west are negated by Bethge's endeavors to do the opposite. So the arena remains the same: twelve hundred meters above the front lines over a scarred expanse of ruins sewn with pockmarks and shell holes, above a transparent little wood, the leafless trunks of which are stuck like toothpicks into the ground broken and slanted.

The R.E. again goes over on its back. Bethge is behind him firing and firing. Then a brownish object detaches itself from the fuselage of the enemy aircraft: the observer. With his machine-guns, which he still firmly clings to even while falling, he disappears into the depths. If the pilot were to land now, Bethge would not fire another shot. Instead he tries to swing around behind the German in a foolhardy turn. Since he does not succeed in doing so against the adroit single-seater, he attempts to escape by diving for his own lines. Now his fate is sealed.

Suddenly he rears up and stands vertical and motionless on his tail for several seconds. As the victor speeds away above him he slowly tilts towards the side and stands on his nose. The projecting ends of the right wing break off, pieces of the aileron fly away and flutter through the air like little glittering white pieces of paper. With racing speed, without levelling out, the fuselage slams into the field of craters.

After landing I shake Bethge's hand. He wipes the sweat from his forehead and points at his machine. It has twenty hits, two of them in the main spar. A grazing shot splintered the left lens of his goggles. Because he was just aiming and closed that eye he came away with only an insignificant wound on his eyelid. He apologizes for poking his nose into my affairs. "But it was impolite of you to insist on seeing more than your own Staffel commander."(15)

As Otto Fuchs later remarked regarding this aerial combat, it was one of the most difficult ones he had ever been a witness.

In the second half of October Bethge was ordered to Berlin. There he fell sick with a mysterious virus. When he returned to the Staffel after about two weeks at the end of October, Otto Fuchs noticed a change in his fatherly friend:

Above: From left, Kurt Katzenstein and Douglas Schnorr enjoy the sun on an autumn day on the veranda of the readiness hut.

Now he is finally standing before me in person – small, compact, full of energy – and scrutinizes me through his large round glasses with well-meaning concern in his eyes. Not as a superior and Staffel commander, not merely as a comrade, but as an older brother and friend.

And yet, for all the joy of seeing him again, I experience a mild fright. One can see in him signs of the illness he had overcome. The features of his face have become sharper. The skin stretched over his cheekbones is indeed tanned, but the paleness beneath shimmers through. The natural lively cheerfulness of before has given way to a deep seriousness which even the occasional emergence of his Berlin wit and his little ironic remarks are unable to obscure. And that is not from the flu!(16)

At the end of the month the order arrived for Otto Fuchs to transfer to Bavarian Flieger-Ersatz-Abteilung 1b at Schleissheim for the assembling of Royal Bavarian Jagdstaffel 77. This transfer had already been announced in writing weeks before, about which Otto Fuchs reported:

One morning when I am still lying in bed and watching the thick snowflakes dancing past the

Above: Paul Erbguth in flight clothing. (Erbguth family)

window Bethge pays me a visit. He has thrown his fur coat over his pajamas and sits on the edge of the bed.

"I have an announcement for you which will certainly bring you great joy," he says. But instead of continuing, he presses his folded hands between his knees and rests his gaze contemplatively on his mahogany red slippers.

"Wonderful slippers you have there ... Lined with camel hair?"

"A good deal at one-fifty." He draws a leg up and hands me the object of my admiration for closer inspection...

"I really don't know how I should break it to you..."

"A telegram from home? My brother. Has something...?"

He shakes his head. "It has to do with you. You seem to have strange notions of joy."

"Certainly the King of England will not want to award me the Order of the Garter? Or am I to be made a head shorter by the end of the day?"

"The southern German lackeys came up with the glorious idea of forming their own colorfast Jagdstaffeln: blue and white, green and white, and so on."

"I don't get it. How's that again?"

"You really are a strange lot, you Bavarians. Say hello for me to the members of your tribe in Munich when you report there on December first..."

"Damned filthy mob!" I jump straight up out of bed and fling Bethge's slipper at the door.

"Take it easy... I have already telegraphed."

"They can do what they want. I'm not going."

"A short letter to the Feldflugchef is already lying ready on my desk. I know him from earlier on. You will stay for the time being. Paul Erbguth, who is also affected, is taking to a Saxon Staffel.

"A Saxon Staffel... That is a Staffel in which all the members speak Saxon dialect?"

"Apparently."

"Hip, hip, hurrah!"

"Fine. I just wanted to get your approval before sending off my note of protest."

"Approval! Anything legal! If you don't fight for me like a lioness for her cub then I will never look at you again!"

"Apparently you don't lack the necessary imagination. I'm not at all surprised, coming from a citizen of Bavaria." I reach for something handy to throw at him. But before I can find anything he springs to the door and sticks his tongue out at me.[17]

This account by Otto Fuchs as well makes clear how informal the relations were between the Staffelführer and the pilots of Jagdstaffel 30, and not only there was this true. In reality, the popular image of a stiff "Prussian militarism" did not exist at all in the Jagdstaffeln. With his petition, Hans Bethge merely delayed somewhat the transfer of Otto Fuchs, but he could not prevent it.

The reason for the establishment of regional Jagdstaffeln was the following: The German Empire consisted of a number of kingdoms, principalities, and duchies. After the Jagdstaffeln had been exclusively "Royal Prussian," the kingdoms of Bavaria, Saxony, and Württemberg demanded their own "Royal Bavarian," "Royal Saxon," or "Royal Württembergian" Jagdstaffeln. Also due to the increasing war weariness the wish was granted, and in the summer of 1917 a series of Jagdstaffeln were accordingly "redesignated."

With the establishment of new units, corresponding "regional" contingents were to be introduced via the site of assembly. If one had left it at that, then this could be considered a mere note in the margins. But the "strategists" in the regional capitals of Munich, Dresden, and Stuttgart had seen

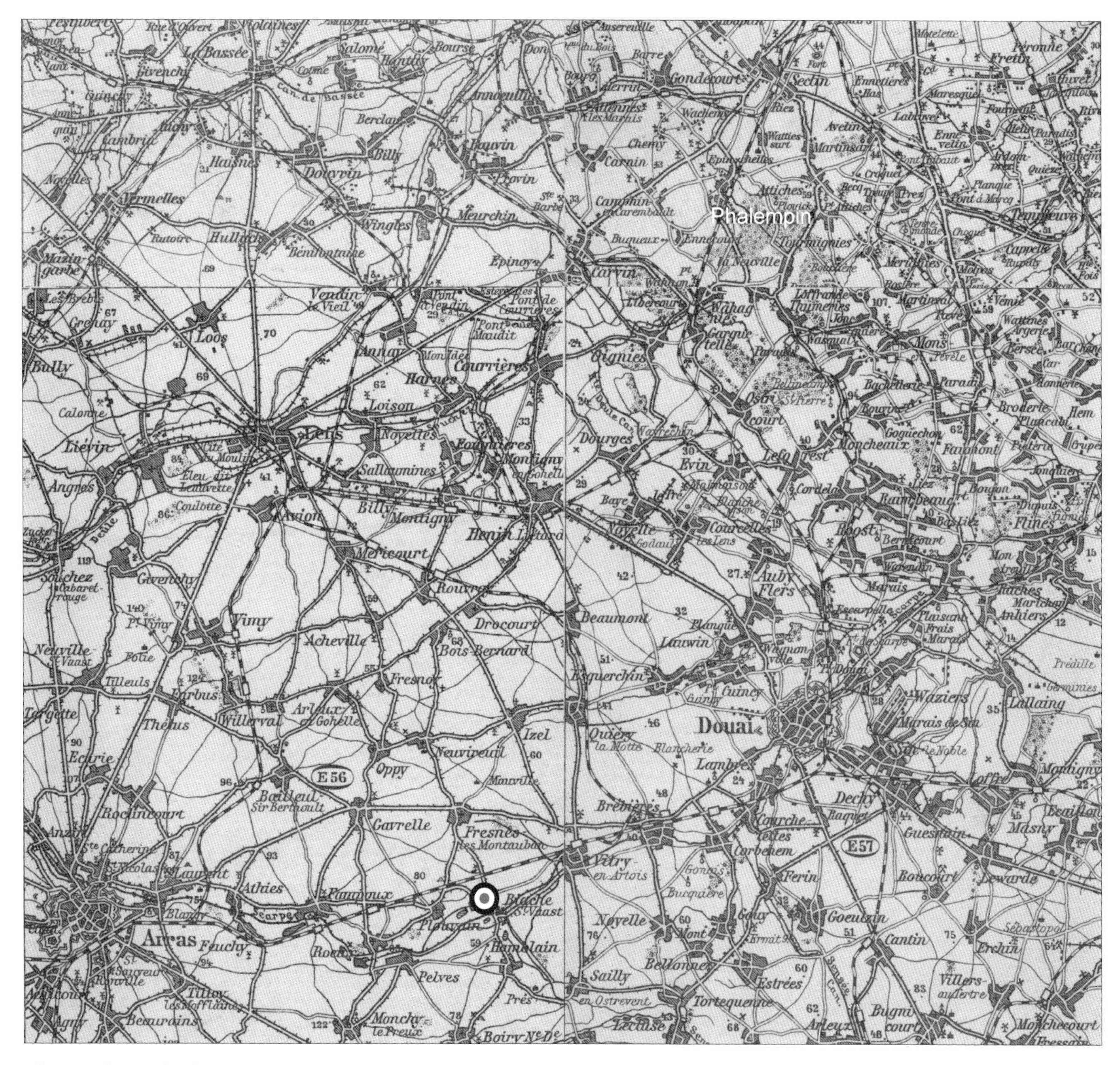

Above: The cockade on the map shows the location of Oblt. Hans Bethge's victory on October 31, 1917 against the brave English crew: Lt. W. L. O. Parker and 1/AM H. L. Postons, 13 Squadron RFC.

to it that only Bavarian, Saxon, and Württemberg pilots were allowed to serve in these units. A curious side note is that the region of origin was not the deciding factor, but rather the matter of with which of the four royal armies one had (by chance) joined up. Thus the Staffel commander whom Otto Fuchs met at Royal Bavarian Jagdstaffel 77 was a "Bavarian" from Lübeck - at that time a city-state within the region of Holstein, in the north of Germany - who during his period of study in Munich joined as a "one-year volunteer" (the normal period of military service at that time lasted two years) with the Bavarian Army.

The creation of regional Jagdstaffeln was criticized especially by many authors in the 1930s, who went so far as to say that "well-functioning, successful Staffeln were torn apart to the detriment of the combat effectiveness of these units," which assertion fit in very well with the so-called "stab-in-the-back legend."

As frustrating as such transfers were for the individual pilots, the formation of these regional units changed nothing with respect to the basic scheme for establishing new Jagdstaffeln which had already been practiced with success. For the leadership of a newly established Jagdstaffel, a

Dienststelle: Königl. Preuß. Jagdstaffel 30

(Standort): Flughafen, den 31. Oktober 1917.

Tgb. No. 7192/17

Personal-Bestandsnachweisung.

Veränderungen zur Kriegsrangliste. (Nur für Offlieg. auszufüllen.)
Abgang: (wohin?), Beförderung, Patentverleihung, Umschulung, Prüfung.
Bei Zugang: (woher?) Kriegsrangliste auf besonderem Bogen beifügen.

(Unterschrift und Dienstgrad): Bethge

Oberleutnant und Staffelführer.

Personal roster of Jagdstaffel 30, October 1917, part 1
Bayerisches Hauptstaatsarchiv Abt. 4, Iluft-Band 206

Flugzeugführer:

Lfde. Nr.	Dienstgrad	Name	Geburtsdatum	Patent	Truppenteil	Zugehörig oder kommandiert	Abzeichen seit	Bei der Dienststelle seit	Woher zur Dienststelle gekommen	Bemerkungen
1	Oblt	Bethge Hans	6.12.90	6.6.16	Leib-Regt. 117	Jasta 30	8.12.15	20.I.17	Jasta 1	Staffelführer
2	"	Preissler Hans	16.7.90	22.3.16	H.R. 21	"		23.10.17	Jastaschule 1	
3	Lt.d.Res.	Schnorr Wiegand	15.9.90	8.9.15	Fr. Abtlg. 19	"	8.6.16	16.I.17	Fa 41	Offz. z. b. Verw.
4	"	Erbguth Paul	24.11.94	22.12.14	Landw. I.R. 107	"	9.5.17	8.III.17	Schusta 21	
5	Lt.	v.d. Marwitz Hans Georg	7.8.93	22.III.15	Ul. Regt. 16	"	31.I.17	18.4.17	Jastaschule I	
6	Lt.d.Res	Fuchs Otto	7.III.97	17.12.15	Feld Art. Regt. 12	"	15.6.17	7.6.17	A.Fl.P.6	
7	Lt.	Welz Paul	24.I.96	21.10.14	I.R. 27	"	2.12.15	25.6.17	R.G.1	
8	Lt.	Frh. v.d. Horst Rudolf	28.6.96	17.12.15	2 Garde Ul. Regt.	"	13.6.17	23.8.17	A.Fl.P.6	
9	Lt.d.Res	Katzenstein Kurt	27.II.95	30.7.17	Fliegertruppe	"	11.5.17	11.8.17	"	
10	Lt.d.Res	Holthusen Hans	13.4.94	6.11.1915	Res. I.R. 90	"	9.6.17	6.10.17	Jastaschule 1	
11	Vzfw.	Oberländer Hans	3.9.94	—	Fa 6	"	17.5.17	17.5.17	A.Fl.P.6	
12	"	Heiligers Josef	23.4.94	—	Fa 7	"	20.I.17	20.I.17	Jasta 4	
13	Uffz.	Funk Josef	9.12.96	—	Fa 10	"	23.5.17	23.5.17	A.Fl.P.6	
14	"	Liebert Emil	16.6.85	—		"	3.6.17	8.9.17	Flg. Abtlg. (A) 244	

Personal roster of Jagdstaffel 30, October 1917, part 2
Bayerisches Hauptstaatsarchiv Abt. 4, Iluft-Band 206

Above: Hans Bethge drives the motorcycle which belongs to the Staffel's fleet of vehicles. (Stiftung Deutsches Technikmuseum Berlin/Historisches Archiv)

pilot officer with frontline experience from an already existing Staffel was chosen, who was named Staffelführer. Pilots from other existing Staffeln who likewise had frontline flying experience were allotted to him as Ketten commanders (the new Staffel commander could in part even select them himself). These three to four "old hands" were to serve as teachers and supporters for the remaining, as yet inexperienced, pilots of the new Staffel during operational patrols and guarantee that the new unit attained the required measure of fighting capability as quickly as possible.

This method was also employed when at the end of 1917 the 40 Jagdstaffeln of the so-called "Amerikaprogramm" were established. That is to say, the transfer of Otto Fuchs to another Staffel would have occurred anyway, since as an experienced and successful Ketten leader he was predestined for it. The creation of the Royal Württemberg, Royal Saxon, and Royal Bavarian Jagdstaffeln, besides the Royal Prussian Jagdstaffeln, merely led to the side effect of a concentration of personnel from various regions from amongst the existing personnel, which possibly led to a better "cohesion" within the individual Staffeln in isolated instances than would have been the case with "mixed" personnel.

It remains to be added that towards the end of the month six Pfalz D III fighter planes arrived at the Staffel for operational testing. It is fairly certain that the background of this allocation for the area of Kofl 6 was the same as for the nearly simultaneous testing by Jagdstaffel 24 (Kofl 4). There, according to its war diary, the Staffel was to try out combined operations with two aircraft types. According to the still extant Jagdstaffel 24 progress report, the combined operation of Albatros D V's and Pfalz D III's was not possible.

At Jagdstaffel 30, on the other hand, it does not appear to have gone that far, since the pilots regarded the Pfalz D III with scepticism, and hardly a pilot with whom the authors spoke had anything positive to say about this airplane. For example, Hans Holthusen completed only a single test-flight of 10 minutes with one of these Pfalz D III's (D.4087/17) on October 24, 1917.[18]

Markings and Paint Schemes of the Aircraft

After the crash landing with his red "F" at the beginning of September 1917, Otto Fuchs received a new Albatros D V. This airplane again received a green fuselage. In order to once more distinguish

Above: Pilots after visiting a French coal mine. From left: French mining engineer (host), Rudolf von der Horst, Douglas Schnorr, Hans Bethge, Hans-Georg von der Marwitz, and a further guest. The odd-looking headgear is the protective helmet customarily used at that time by French miners.

his green from the "grass green" of Paul Erbguth, he again mixed up a special green, which however was tinged with blue this time. He later characterized this as "verdigris green."

At about the same time he had discovered the silhouette of a fox in a magazine. He liked the idea of representing his name in this manner on his airplane ["Fuchs" being the German word for "fox"]. He sketched the drawing on a large piece of paper and transferred it to two thin plywood boards. His mechanics sawed out two foxes from the plywood and brushed them with black, waterproof paint. In addition, as presents his mechanics sawed out two Gallic cocks and painted them red, which the fox symbolically chased.

In order that the fox would stand out well from the green fuselage, a white oval was painted on the fuselage and surrounded with a black border. Then the foxes and the cocks were screwed onto it. In the report of his and Hans Bethge's aerial combat on October 31, 1917 he mentions the "verdigris green fuselage" of his Albatros D V (Profile 25). When Otto Fuchs was transferred to Jagdstaffel 77, he took the foxes and the cocks along with him and at the new Staffel screwed them onto the fuselage of his new machine, an Albatros D III (OAW).[19]

The "red F" later came back to the Staffel after a general overhaul, but was no longer flown by Otto Fuchs.

With the assistance of his mechanics, Lt. von der Marwitz had painted his new Albatros D V wine red, concerning which Otto Fuchs remembered (Profile 26):

When Hans-Georg received the new crate, he disappeared the next day into the hangar with two bottles of red wine. Later I found him there together with his mechanics. He had tied on a green gardener's apron, and in his right hand he held a big brush and in the left a glass of red wine. That was Hans-Georg all over.[20]

For Kurt Katzenstein Otto Fuchs painted a "cat on a stone" as an allusion to his name ["Katze" = cat; "Stein" = stone] on both sides of his airplane. In addition, the tail of this Albatros D V was painted

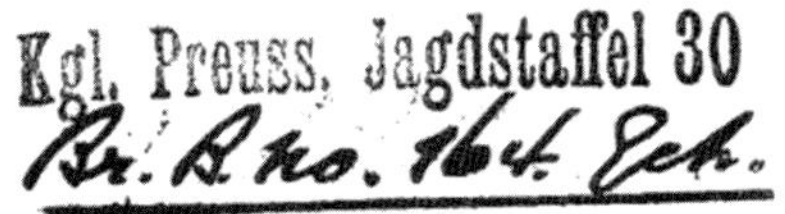
Kgl. Preuss. Jagdstaffel 30
Br. B. no. 164. Geh.

B e u r t e i l u n g

des Lt.d.Res.~~Oscar~~ Fuchs der Kgl.Preuss.Jagdstaffel 30.

Lt.d.R.Fuchs ist seit 5.Juni 1917 als Flugzeugführer bei der Staffel und hat sich bei derselben als Jagdflieger ganz hervorragend bewährt. Aus einer grossen Anzahl von Luftkämpfen hat er drei anerkannte Siege davongetragen. Seine Flugfreudigkeit und sein kühl überlegender Angriffsgeist machten ihn zu einem gesuchten Kettenführer, der ohne Zweifel noch bedeutende Erfolge haben wird.

Trotz seiner Jugend als Offizier und an Lebensalter besitzt er ein sehr sicheres, doch bescheidenes Auftreten, sowie den nötigen Ernst. Er war im Kameradenkreise infolgedessen, sowie seiner tadellosen gesellschaftlichen Formen und seiner reichen Belesenheit ein recht beliebter und geachteter Kamerad.-

Flughafen, 13.Nov.1917.

Bethge

Oberleutnant u.Staffelführer

Above: Assesment of Otto Fuchs by Oblt. Hans Bethge Bayerisches Hauptstaatsarchiv Abt. 4, Personalakte Otto Fuchs.

Assessment of Ltn. der Reserve Otto Fuchs

Since June 5th 1917, Ltn. der Res. Otto Fuchs has been a pilot with my unit and has proved himself as an excellent fighter pilot. From a large number of air combats he achieved three confirmed victories. His eagerness to fly and his cool and prudent spirit to attack made him a desired Ketten leader, who will no doubt have important success.

Despite his youth as an Officer he has a confident but modest behaviour and the needed seriousness.

Because of this, his excellent manners, and scholarship he is well liked and respected among his comrades.

Airfield 13th November 1917 Bethge

black. Considering the fact that Otto Fuchs was a landscape painter and not an animal painter, the cat looked, according to his statement, a little strange, but "Kurt was very happy." (Profile 27)

Rudolf von der Horst "had to" have the fuselage of his Albatros D V painted white. He had a great passion for candy, especially sugar eggs, which his mother regularly sent to him at the front. Even while flying missions he nibbled on these candies. This was of course the occasion for many a joke, and the other pilots decided unanimously that his aircraft had to be painted white. Because the white paint was applied to the plywood fuselage without a coat of primer, the white gleamed with a bit of a cream color. It is no wonder that his airplane afterwards received the name "the sugar egg." (Profile 28)

On October 24, 1917 Lt. Hans Holthusen had taken over "his" Albatros D V D.2126/17. He had his aircraft painted in the colors of his home town of Hamburg, white with a red spiral band. The machine therefore had the nickname "the corkscrew (Profile 29)."[21]

Lt. Weltz' Albatros D V once more had a "sky blue" fuselage and that of Uffz. Emil Liebert had a dark blue fuselage (Profile 30).

On the Albatros D V of Uffz. Josef Heiligers the

D	Albatros	D.V 2051/17	Albatros	D.V 1023/17	Wegen starker Beschädigung durch Zusammenstoß mit einem anderen Flugzeug	A. Fl. P. 6
	"	" 2203/17	"	" 2051/17	Im Luftkampf abgeschossen	restlos verbraucht.
	"	" 2197/17	"	" 1016/17	Befehl Kofl. 6 Tgb. Nr. 3597 VI	A. Fl. P. 6
	"	" 2140/17	"	" 1012/17	" " "	" "
	"	" 2126/17	"	" 2047/17	Im Luftkampf stark beschädigt.	" "
	"	" 2199/17	"	D.III 2304/16	ausgeflogen	" "
	"	" 4422/17	"	" 2132/16	"	" "
	"	" 2150/17	"	" 2054/16	"	" "
E	"	" 2186/17	"	" 2141/16	"	" "
D	"	" 2205/17	"	" 767/17	Stark beschädigt infolge Motordefekt	" "
	"	" 2207/17	"	" 760/17	" " durch Überschlag bei der Landung	" "
	"	" 2177/17	"	D.V 2191/17	über dem Feinde verloren	
	"	" 2198/17				
	"	" 4420/17				
	"	D.III 605/17				
	"	" 791/17				
	"	" 799/17				
	"	" 2306/16				

Above: Aircraft inventory of Jagdstaffel 30, October 1917 Bayerisches Hauptstaatsarchiv Abt. 4, Iluft-Band 206

"X" was displayed again, only this time it was red without a border (Profile 31).

Oblt. Bethge had his Albatros D V painted mouse gray with a broad white stripe. The reason was that he had first allowed his pilots to choose the paint schemes of their aircraft and in the end no other color was left over for him. In addition, he had a preference for the color gray. He loved the melancholy mood of gray, foggy days in northern France on which he took long walks. Accordingly, the aircraft was called "Bethge's Mouse."[22] (Profile 32)

Right: Sopwith Camel B6314 of Lt. Winter at Armee-Flug-Park 6. The British national insignia on the rudder was retouched on the photo.

4. November 1917 – February 1918

Above: The Pfalz D III 4079/17 and the wine red Albatros D V of Lt. von der Marwitz in the middle in November 1917. On the fuselage and on the elevators can be seen the orange-colored, black-bordered Staffel marking. (A. Imrie)

Farewell to the Albatros

At the beginning of the month of November 1917 the bad weather prevented larger deployments of flying units in the Army sector.

On one of his last days with Jagdstaffel 30, a gray and rainy November day which hardly allowed any aerial activity, Otto Fuchs took a long walk with Hans Bethge, during which he discovered the reason for the change in him. Hans Bethge brought up his experiences back home, about which Otto Fuchs related:

Hans Bethge had only stayed a few days in the capital, where the pleasure industry was thriving no less than the black market. Anyone who was not a profiteer went hungry—and just how many were literally starving could not be determined. He had been invited to come out to the country by an uncle, the owner of an estate, where of course nothing was lacking as far as food was concerned. However, his two sons had fallen in battle and their mother had become melancholy.[1]

In the gloomy mood of his farewell Otto Fuchs suddenly asked Hans Bethge whether he believed Germany could still win the war, whereupon the latter answered:

There's no point in fooling oneself. We're not the only two who know it. The Supreme Army Command knows it just as well and certainly has known it much longer than we... we all too credulous fools! The fact that they don't draw the conclusions from that, don't have the courage to confess that they had miscalculated, attempt to keep the people at it with lies and create illusions which no one buys anymore who possesses even a spark of common sense – that is scandalous. That is a crime against our people. We didn't deserve that. Instead of finally calling an end to the senseless bloodletting, they always plan new offensives with ever less adequate means, send thousands upon thousands to their deaths, simply in order to be able to claim afterwards that they would have held out till victory was achieved, but the homefront had

Above: With four aerial victories, Vzfw. Josef Heiligers was one of the successful pilots of the Staffel.

stabbed them in the back and finally they also left the frontline in the lurch.

Oh, of what concern to me are the cowardly personal motives of the higher and highest ranks of power? We have enough to deal with amongst our own... I do, anyway![2]

Otto Fuchs was dismayed at this passionate outburst of despair and the anger of his friend and Staffelführer; he had never seen him like this. Otto Fuchs had no idea that this would be the last long conversation he would ever have with Hans Bethge, but still that night he wrote down a detailed report from memory concerning it.

One or two days later Hans Bethge brought him personally to Armee-Flug-Park 6. Rarely during his life had Otto Fuchs found a parting so difficult.

On November 8, 1917 the weather improved. Hans Holthusen reported in his logbook about an aerial combat with six Bristol fighters, which however he had to break off due to a jam. At noon on the same day a formation of D.H. 4s from 18 Sqn. RFC bombed first Esquerchin and thereafter the airfield at La Brayelle southwest of Douai. In the end, at the north exit of Brébières the "Englishmen" were attacked by two Albatros D Vs of Jagdstaffel 30 which had taken off in the meantime. While doing so Vizefeldwebel Hans Oberländer was able to force down one bomber towards the southeast, causing it to make a forced landing in the vicinity of the mine at Monchecourt around 13:00, during which it flipped over and caught fire. The crew was taken prisoner.

The patrols during the following days were for the most part uneventful, apart from the aircraft regularly being fired upon by British flak.

On November 10, 1917 Hans Bethge, whose virus had still not entirely worn off, went on a four-week furlough which had long been approved. The acting leadership was taken over by the longest-serving officer of the Staffel, Oblt. Kurt Preissler. In the air, however, the Staffel was led by Hans-Georg von der Marwitz, to whom Paul Erbguth had given precedence.

After thick mist had prevailed during the early hours of November 19, 1917, airplanes from the Staffel had taken off in the morning towards the

Below: The wreckage of the black Albatros D V of Vzfw. Josef Heiligers. The airplane crashed over the railway triangle of Ostricourt due to an engine fire. (R. Zankl)

Right: Albatros D V 4420/17 as a school aircraft with Flieger-Ersatz-Abteilung 13. The orange lozenge indicates that it is a former machine of Jagdstaffel 30. It is perhaps the light blue Albatros D V of Lt. Weltz.

southwest in the direction of Méricourt and Oppy, as enemy aerial activity had been reported in this sector of the front. Since no enemy aircraft were encountered there, the Staffel turned towards the north, passing by the town of Lens and attacking enemy trenches with machine-gun fire from a low height near Hulluch. Upon request from the front, the Kette of Hans Holthusen took off anew, but this time as well there was no aerial combat.

On November 20, 1917 began the so-called "tank battle of Cambrai," involving a large British attack in the sector of the German 2nd Army with its thrust directed towards the town of Cambrai. It received its name from the massive deployment of British armored vehicles which even at that time were called "tanks." In order to support the flying units of Kofl 2 against the superior Allied forces, the Staffeln of the neighboring German 6th Army to the north were deployed in the Cambrai area if the situation in the air required it.

The Staffel took off amongst gusty winds for the first time for such a deployment with the German 2nd Army. Even on the way to the deployment area, flames suddenly shot out of the engine of the black Albatros D V of Josef Heiligers above the railway triangle of Ostricourt. The machine caught fire and crashed in flames around 13:47. Josef Heiligers, one of the best pilots of the Staffel, burned up in the wreckage.

A day later Jagdgruppe "Nord" (North) of the 6th German Army was formed from Jagdstaffeln 18 and 30. The leadership of the Jagdgruppe was assigned to the commander of Jagdstaffel 30. Since Oblt. Bethge was still on leave, the task at first fell to his stand-in, Oblt. Preissler. The latter took over the coordination of the deployment of both Staffeln, while Lt. von der Marwitz continued to lead the Staffel or the group formation in the air.

Above: The grave of Uffz. Wilhelm Foege, who fell on December 22, 1917, in the cemetery of Phalempin.

However, aerial activity on both sides of the front was slight due to the bad weather, as a result of which the Jagdgruppe found little action. Accordingly, in the entire month of November only 14 British aircraft in the area of the 6th German Army were shot down or forced to land. Of these,

Right: New Year's Eve celebration of the Staffel, 1917. From left: Hans Oberländer, Hans Holthusen (blurred), Hans Bethge, Rudolf von der Horst, Erich Kaus, Kurt Preissler, Paul Erbguth, Lt. Seewald, and Reinhold Maier.

Left: The Sopwith Camel of 10 Sqn. RNAS forced to land by Vzfw. Emil Liebert on January 3, 1918. The nose of the machine was striped white and red.

Right: The Sopwith Camel N6351 from the same squadron forced to land on the same day by Vzfw. Oberländer. This machine had a blue and white striped nose.

Above: From left: Vzfw. Oberländer, Lt. Seewald, Lt. Holthusen, and Lt. von der Horst in December 1917 on the airfield.

only two were downed in aerial combat, while the rest fell victim to groundfire. In the previous month 36 enemy aircraft were shot down.

After the British offensive had not brought the hoped-for breakthrough through the German lines into the rear area, the German counterattack began on November 30, 1917, completely surprising the British. Through this attack almost the entire territory just lost was able to be recaptured, but by and large it likewise did not lead to a change in the overall military situation. Once again tens of thousands of soldiers had laid down their lives in the end for nothing.

On December 2, 1917 Lt. Erich Kaus arrived at Jagdstaffel 30. He was an experienced pilot who had already begun his pilot training in February 1916. On August 21, 1916 he was transferred to Kampfstaffel 8 in Kampfgeschwader 2. After he had proven himself there as a good and brave pilot, he was transferred on March 28, 1917 to the newly established Jagdstaffel 31. There he and two other pilots had a bad falling out with their Staffelführer (an Oberleutnant), who was of course displeased with the massive criticism of the three "mutineers" and had them transferred on the spot to their reserve units. This disciplinary transfer occurred on August 9, 1917 (incidentally, the Herr Oberleutnant himself was relieved of his post in September and likewise transferred).

While with Field Artillery Regiment 55, Erich Kaus wrote several requests to be employed as a fighter pilot, of which Hans Bethge learned through Manfred Freiherr von Richthofen. Kaus' courage, combined with his frontline experience, impressed Oblt. Bethge and he brought about his transfer to Jagdstaffel 30.[3] One of the other "mutineers" – Leutnant Richard Wenzl – was by the way fetched by von Richthofen for his Jagdgeschwader I and there, for his successes as a fighter pilot, he was awarded the Knight's Cross of the Royal House Order of Hohenzollern.

On the morning of December 2, 1917 Jagdstaffel 30 received the order to become engaged in the area of the German 2nd Army in support of the German counteroffensive. On this day Hans Holthusen was flying an old Albatros D III since "his" D V was not ready for action. When he returned from another patrol on the same afternoon with this Albatros D III he was caught by a gust from the side while landing. The aircraft flipped over and was wrecked in the process. However, he remained uninjured, apart from a few bruises. During the following days he flew patrols in the Albatros D V of one of his comrades.

On December 5, 1917, a day with clear weather, the Kette of Hans Holthusen was deployed in the area of the German 6th Army, as well as the 2nd Army. Over Cambrai there was an aerial combat with five of ten S.E.5s sighted, and then later there followed a combat with 8 Sopwiths near Vacquerie. Both aerial combats were without result.

On December 6, 1917 the first Pfalz D IIIa's were transferred to the Staffel. Already in October 1917 the pilots had had an opportunity to test some Pfalz D IIIs briefly allocated to them and were not at all enthusiastic about this aircraft. The slightly improved Pfalz D IIIa's now assigned to them were considered difficult to fly, unpleasant to land, and fairly trouble-prone. If the Albatros D V and D Va were inferior to the latest Allied fighters and even many two-seaters, then this was even more true of the Pfalz D IIIa despite its nominally comparable flight performance. Therefore every pilot of the Staffel tried to keep his Albatros D V as long as possible. But in the middle of February 1918 the last Albatros fighters had to be given up to the Flugpark by order of the Kofl.

The background to the allocation of the Pfalz D IIIs was that the Albatros-Werke could no longer meet the needs of the Jagdstaffeln due to capacity bottlenecks. In addition, the Bavarian government had an (understandable) interest in employing the Bavarian Pfalz-Flugzeugwerke in Speyer to full capacity. Therefore pressure was apparently applied on the political track so that, besides the Bavarian Jagdstaffeln, Prussian, Saxon, and Württemberg Staffeln were also equipped with Pfalz aircraft.

On December 10, 1917 Oblt. Hans Bethge

Above: The wreck of the Albatros D V of Uffz. Emil Liebert after his fatal crash on January 3, 1918. (R. Zankl)

returned from leave and once again took over command of the Staffel. Two days later Jagdgruppe "Nord" was expanded to include Jagdstaffel 29.

However, now as before, bad weather restricted patrol activity. Thus the logbook of Hans Holthusen records only eight patrols in the period of December 7 to December 18, 1917, during which there was not a single aerial combat.

On December 22, 1917 around 14:20 there was an aerial combat with Spads of 19 Squadron R.F.C., during which Sergt. Wilhelm Foege was badly wounded by a hit in the lung. He succeeded in making an emergency landing in his machine, but he died shortly thereafter at the spot where he landed.

On December 28, 1917 Lt. Paul Erbguth received the order to transfer as Staffelführer to the newly established Royal Saxon Jagdstaffel 54. Shortly before his transfer the report arrived at the Staffel that he had been awarded the highest Saxon decoration, the Royal Saxon Military St. Heinrich Order. The official award document reads:

Erbguth, Paul, Lt. d.R. in Royal Saxon Territorial Reserve Infantry Regiment 107, of Royal Prussian Jagdstaffel 30; born on 20.11.1891 in Reichenbach (Vogtland); in peacetime student of natural sciences technology, awarded on 2.12.1917: After Lt. Erbguth was badly wounded in the leg with Terr. Res. Inf. Reg. 107 on 30.04.1915 he reported, barely recovered, to the air service. Belonging to the latter since 23.4.1916, he had revealed himself to be an exceptionally brave and prudent pilot in a great number of aerial combats. On 21.9.1917 he shot down his second enemy aircraft over our lines.[4]

Up till the end of the year the winter weather continued to severely limit flight operations and therewith gave the Staffel a relatively peaceful Christmas. It remains to be added that on December 30th Lt. Reinhold Maier arrived at the Staffel.

On January 3, 1918 aerial activity was revived on both sides of the front. At this time there was a determined dogfight between machines of Jagdstaffel 30 and Sopwith Camels of 10 Sqn. RNAS. Lt. Kaus, along with Vzfw. Oberländer, attacked an opponent with a blue- and white-striped motor cowling and forced the airplane to land. Concerning this he reported:

I myself was able to force the Englishman to land (the Englishman flipped over). Since he did not leave the aircraft, I flew up to it once more and fired some shots next to the airplane. Thereafter he climbed out. However, the victory was not confirmed for me, as Oberländer maintained that he had been able to shoot him down.[5]

Uffz. Emil Liebert had attacked a second machine from the enemy formation with a red and white striped cowling, which went down under the fire of his machine-guns and crashed from a low height. The two British aircraft came down only a few meters away from each other along the Meurchin – Provin rail line. Erich Kaus and Emil Liebert landed

Above: From left, Lt. Holthusen, Lt. von der Marwitz, Oblt. Bethge, Lt. Erbguth, and Lt. Kaus in front of Bethge's new Pfalz D IIIa D.4203/17.

in the vicinity of the downed enemy airplanes and were immediately surrounded by German soldiers, who cheered them. While Lt. Kaus took off afterwards with no problems, while taking off Uffz. Liebert collided with the safety rope of a captive balloon of the Bavarian 23rd Balloon Company and crashed fatally.

On January 6th Lt. Hans Holthusen ferried Albatros D V D.2203/17 over to Aflup 6. On the same day he flew Pfalz D IIIa D.5891/17 from there to his old Flg. Abt. 18 for a visit and on the next day flew back to Jagdstaffel 30, which certainly did not represent standard procedure.

On January 13th he took off in the morning with the same Pfalz for a ten minute test flight, and thereafter made a patrol with his Kette. During this action he first attacked a Bristol Fighter without result. Afterwards there followed an aerial combat with four Sopwiths. In the course of this fight the auxiliary spar and the aileron broke. He had to break off the combat and return to the airfield. It is no wonder that during his subsequent missions he went back once more to his accustomed Albatros D V D.2126/17.[6]

Only on January 24, 1918 did he take off with Pfalz D IIIa D.5891/17, which had in the meantime been repaired, for a flight at the front. The patrol lasted 70 minutes, in the course of which he attacked a D.H. 4, which, however, was able to escape.

The next day Hans Holthusen took off with several airplanes from the Staffel for the front, after lively enemy aerial activity had been reported from there on account of the fine weather. There was an aerial combat with five D.H. 4s and a Bristol Fighter, which proceeded without result. Also the further deployments of the Staffel up to the end of the month remained without success.

On January 28, 1918 a Kette of the Staffel was attacked by S.E.5a fighters of 40 Sqn. RFC between Drocourt and Beaumont. In the process 2/Lt. Hutton succeeded in placing himself behind the machine of Lt. Reinhold Maier and opening fire. The British pilot saw his burst enter the fuselage of the German airplane, which thereafter went into a vertical dive trailing a bluish stream of smoke. Because the radiator of his S.E.5a was damaged 2/Lt. Hutton, who at the same time was hard pressed by two other aircraft of Jagdstaffel 30, had to turn away. Reinhold Meier was able to make an emergency landing on the

Above: The Pfalz D IIIa 4203/17 of Oblt. Bethge at the beginning of 1918. The rudder is orange with a black outline.

German side and was brought wounded to a military hospital.

In the course of the month Lt. Katzenstein and Lt. Seewald had to leave the Staffel. Kurt Katzenstein was transferred to Aflup. F to see frontline duty with Jagdstaffel 55 (1F) in Palestine and Lt. Seewald to Jagdstaffel 54 by Paul Erbguth. In their stead two novices, Lt. Morhau and Lt. Wendel Bastgen, arrived at the Staffel.

On February 7, 1918 the composition of Jagdgruppe "Nord" of the German 6th Army was changed once more. It now consisted of Jagdstaffel 29 at Bellincamps airfield, Jagdstaffel 30 at Phalempin, and the newly established Jagdstaffel 52, which had moved to the airfield at Bersée.

The aerial combat activity in the area of the German 6th Army in the first half of February was slight due to weather conditions. Only on February 9, 1918 was increased enemy flight activity reported in the area of Group "Souchez" in the southern army sector. The logbook of Hans Holthusen as well records only four flights at the front, all of them without result.

Eight days later, on February 15, 1918, Oblt. Hans-Joachim Buddecke, recipient of the Order "Pour le Mérite," arrived at the Staffel. He was one of the first fighter pilots who had achieved his initial successes in air combat as a Fokker pilot in 1915. Assigned, with interruptions, to the Dardanelles front in Turkey since December 1915, he made a name for himself there as an exceptional and successful airman. Back on the Western Front, he was to take over Jagdstaffel 18 from his friend Oblt. Rudolf Berthold, who was wounded. To begin with, he was ordered to Jagdstaffel 30 as acting Staffelführer in order to receive a brief introduction to the conditions of aerial combat in the western theater of war, which were entirely new to him.

On the next day Lt. Wendel Bastgen, who had only been with the Staffel for three weeks, was forced to land during aerial combat with S.E.5a fighters of 1 Sqn. RFC over enemy territory southeast of Bailleul and fell into captivity. A report about this by 2/Lt. Percy Jack Clayson is available:

I saw an enemy aircraft at an altitude of about 4000 feet over Mont Rouge. Fired upon by anti-aircraft fire, I climbed above him, whereupon his machine went down in the direction of Bailleul. I dived down after him and opened fire at a short distance. After three bursts the enemy plane went down and flipped over in a ploughed field. I saw the pilot walking

Above: Erich Kaus in the pilot's seat of Bethge's reserve aircraft.

around his machine.[7]

On the morning of the same day Hans Holthusen had an aerial combat with enemy aircraft which he described as "Sopwiths." While doing so, the auxiliary spar of his Pfalz D IIIa broke yet again. In addition, a cabane cable was torn. Again he had no choice but to break off the combat with a damaged machine and fly back to the airfield. The following patrols were flown with Bethge's reserve machine, before on February 19, 1918 he took off for his last flight at the front in "his" Albatros D V D.2126/17. The flight lasted 65 minutes, during which there was no contact with the enemy.

Despite generally moderate aerial activity on the part of the British, on February 19, 1918 there was a fight at the front around 14:00 between a Kette of Jagdstaffel 30 and a flight of 80 Sqn. RFC. Oblt. Buddecke was the first to succeed in opening fire on a Sopwith Camel, which thereafter had to make an emergency landing in the British lines. Only five minutes later Oblt. Bethge got hold of another Camel and forced it to land on the German side between Aubers and Holpegarde. Almost simultaneously a third Sopwith crashed in flames under fire from Lt. von der Marwitz over German territory near Marquilles. Hans Holthusen, who also participated in the combat, had to break off due to a stoppage.

On February 21, 1918 Hans Holthusen climbed aboard the Staffel "hack," D.F.W. C.V C.6401/17, in order to fly Lt. Schnorr and his dog "Beppo" to Jagdstaffelschule I at Valenciennes. To the disappointment of Douglas Schnorr and the entire Staffel, he had been transferred as Offizier z. b. V. to Jagdstaffel 38 on the Macedonian front.

The entry in Lt. Holthusen's logbook reads:

Flight from Jasta 30 to Jastaschule I. Flight guests Lt. Schnorr and Beppo, flight time: 20 minutes, altitude 2000 meters.[8]

One of the few days with somewhat good weather was February 26, 1918. Hans Holthusen used this for two flights to his old Flg. Abt. 18 in order to invite two friends to the Staffel. First he picked up Oblt. von Boddien, who was there on a visit, and then his former observer, Lt. Schröder.

For the rest of the month flight activity was again restricted by bad weather and there were only sporadic aerial combats.

Right & Below: The remains of the Pfalz D IIIa of Lt. Maier after his emergency landing on January 28, 1918. The Staffel marking has also been applied to the underside of the horizontal stabilizer surfaces.

Markings and Paint Schemes of the Aircraft

In November 1917 a new Staffel marking was introduced. An orange lozenge outlined in black was painted on both fuselage sides and on the upper and lower surfaces of the horizontal tail.[9] Since Otto Fuchs could not remember such a marking and none of his known aircraft bore such an insignia, and on the other hand the Albatros D V of Vzfw. Heiligers which crashed on November 20, 1917 already had the Staffel marking on the fuselage, it must be assumed that this marking was introduced around or short after November 15th.

The repeated introduction of a Staffel marking was owed not so much to a whim of the pilots, but rather to military necessity. The ever increasing number of Staffeln and additionally their coordinated deployment in Jagdgruppen simply made a unified Staffel marking necessary as an aid in leading a unit. The Staffel markings had to be reported to the Kommandeur der Flieger of the respective army, who passed on this information to the flak units and air observers (which also made the confirmation of victory claims easier). It is therefore very probable that the introduction of the Staffel marking is directly connected to the formation of "Jagdgruppe Nord of the 6th German Army" on November 22, 1917.

It is, however, not entirely certain whether the Albatros D Vs of the Staffel also received the marking. The wine red Albatros D V of Lt. von der Marwitz and the sky blue Albatros D V of Lt. Karl Weltz had the orange-colored lozenge (Profiles 33 and 34). (For example, a photo taken in 1918 at FEA 13 in Bromberg shows Albatros D V D.4420/17 with such a paint job. Since this machine can be proven as part of the Staffel's inventory, this must concern a photo of the former machine of Karl Weltz.)

Uffz. Emil Liebert subsequently had the lozenge applied to his dark blue Albatros D V. (Profile 35) Also the Albatros D V of Vzfw. Josef Heiligers, which had a black fuselage with a white band at the height of the national insignia, bore the lozenge on both sides of the fuselage (Profile 36). As far as Hans Holthusen could remember, his "corkscrew" in the end also bore the lozenge on its fuselage (Profile 37).

Above: Phalempin airfield in February 1918. On the far left an Albatros D V can be seen half-obscured by a Pfalz. The second machine from the right is the Pfalz D IIIa of Lt. Erich Kaus. On the far right side of the picture is the Staffel hack, DFW C.V C.6104/17.

Right: The Pfalz D IIIa of Lt. Rudolf von der Horst in February 1918. His two mechanics are to be seen in front of the machine. In the hangar in the background an Albatros D V can also be seen.

Below: Lt. Rudolf von der Horst zu Hollwinkel in front of his Pfalz D IIIa. (A. Imrie)

Above: Lt. Erich Kaus on a Pfalz D IIIa.

The British G-report about Albatros D V D.4422/17 of Lt. Bastgen, captured on February 16, 1918, mentions as its paint scheme a yellow fuselage with a black line running diagonally, but no orange-colored lozenge. Instead the report speaks of an "insignia" on the tail plane. This may have concerned the lozenge. This allows the supposition that the yellow of Lt. Bastgen's airplane may have been dark, as the difference to the orange-colored lozenge is not specially mentioned (Profile 38).

When the Pfalz D IIIa's arrived at the Staffel in January 1918, the Staffel marking was adopted by the new airplanes. Some of the Pfalz aircraft also had the lozenge on the top surface of the upper wing; here it may have been a matter of the marking of the Kette leaders' machines.

Besides the Staffel marking, Oblt. Bethge's Pfalz D IIIa 4203/17 had the rudder painted orange (Profile 39). The paint scheme of his reserve machine is not documented.

Lt. von der Marwitz as usual had the fuselage of his Pfalz painted wine red. Hans Holthusen adopted the "corkscrew" paint scheme of his Albatros D V for his Pfalz D IIIa (Profile 40).

Rudolf von der Horst had the stern of his Pfalz painted white and edged with a red band – in honor of his original unit, the 2nd Uhlan Guard Regiment, whose uniform had red trim on its caps, collars, and cuffs (Profile 41).

In addition to the lozenge, Lt. Reinhold Maier added a black "M" and had the stern of his Pfalz painted with black and yellow lozenges, which were the colors of the coat of arms of the Kingdom of Württemberg (Profile 42).[11]

Above: Jasta 30 in February 1918. First row, from the left: Lt. Holthusen, Oblt. Preissler, Oblt. Bethge, unknown, Lt. von der Marwitz, Lt. von der Horst. Second row, from the left: Lt. Kaus, Lt. Schnorr, unknown, Oblt. Buddecke, unknown.

Above: The old and the new Offz. z. b. V. of the Staffel on February 21, 1918. From left: the departing Lt. Douglas Schnorr with his dog Beppo, next to him his successor Oblt. Kurt Preissler, and on the right Lt. Hans-Georg von der Marwitz.

Above: The farewell photo of Lt. Douglas Schnorr taken shortly thereafter in front of the Staffel hack. The DFW C V likewise bears the Staffel marking on the fuselage. From left: Oblt. Hans Bethge, Lt. Schnorr, and Lt. Hans Holthusen.

Below: Oblt. Buddecke after returning from an operational flight with Pfalz D IIIa D.5983/17 in conversation with his mechanics. (A. Imrie)

Above: Pilots of the staffel photographed on the 26th February 1918, from left: Lt. Holthusen Uffz. Funk, unknown, Lt. von der Marwitz, Lt. Schröder (visitor from Fl.Abt. 18), Lt. Kaus, Lt. von der Horst, Oblt. Preissler. (Stiftung Deutsches Technikmuseum Berlin/Historisches Archiv)

Lt. Erich Kaus painted the fuselage of his Pfalz D IIIa from behind the engine cowl to the tail in yellow and white, the colors of the Kingdom of Hannover. In this case he used a painting similar to one he had already used as a personal marking on his Albatros D III as a member of Jagdstaffel 31 (Profile 43).[12] Hans Holthusen used the "corkscrew" design of this Albatros D V for his Pfalz D IIIa (Profile 44).

The personal insignia of the Pfalz D IIIa flown by Oblt. Buddecke was characterized by Erich Kaus to Alex Imrie as "the zodiac sign Cancer." Unfortunately no photo is available showing the entire machine. However, one can assume that this Pfalz also had the Staffel marking painted on the horizontal tail surfaces.[13] This allows the conclusion that the pilot of the machine was born under the sign of cancer. In the spring of 1918 there was only one pilot to whom this zodiac sign applied, Uffz. Otto Busch, who was born on July 19, 1892. He arrived at the Staffel on September 24, 1917. On December 8, 1917 he was transferred to Armee-Flug-Park 6 and on December 21 back to Jagdgruppe "Nord". Since the leadership of Jagdgruppe "Nord" lay with Oblt. Bethge, this most probably means that he returned to the Staffel at this time. On February 15, 1918 followed his transfer to Armee-Flug-Park 6. On the same day Oblt. Buddecke was transferred to the Staffel and thus may have taken the orphaned aircraft for his frontline patrols (Profile 45).[14]

A photo also shows a Pfalz D IIIa with a white fuselage and black bands; unfortunately, none of the former members of the unit asked could remember who flew this airplane (Profile 46).

5. March 1918 – June 1918

Above: Oblt. Bethge and Lt. Holthusen in an excited conversation after returning from an operational flight in March 1918.

The Death of Hans Bethge

At the beginning of the month of March 1918 heavy ground mist hindered aerial activity. Only on the 6th and 7th of the month was there fair and dry spring weather which the Staffel could use for operational flights. Thus the "Holthusen" Kette took off for the front twice in the morning and once in the afternoon on March 6th. During the second flight there was an inconclusive combat over Biez Wood with Sopwith Camels and R.E.8s. Two further patrols were flown the next day, during which however there was no contact with the enemy.

On March 8, 1918 Oblt. Buddecke left the Staffel in order to take over the acting command of Jagdstaffel 18; only two days later he was shot down and killed in aerial combat during his first flight with his new Staffel.

Hans Holthusen patrolled with several machines of the Staffel on the morning of the same day along the rail line Phalempin-Douai and near Douai ran into five S.E.5s. These were immediately attacked, but the ensuing combat was without results. After the Pfalz D IIIa's had reassembled, they turned westwards in the direction of Lens, in order to then fly along the front towards the north. Over Biez Wood, east of Neuve Chapelle, they sighted four Bristol Fighters, which were likewise attacked. But this combat as well produced no significant result.

Two days later Hans Holthusen was aloft with several comrades under the leadership of the Staffelführer. The Staffel attacked six D.H.4s near Allennes. In the course of this aerial combat Oblt. Hans Bethge was able to shoot down an opponent at 12:10 and therewith achieved his 20th victory.

Accordingly, on March 12th the following recognition appeared in the army orders:

Above: Four Pfalz D IIIa's in front of the hangar at Phalempin airfield in April 1918. The second machine from the left is Lt. Holthusen's "corkscrew." Next to it on the right is a Pfalz D IIIa with black bands on the fuselage, and at far right is possibly Oblt. Bethge's reserve machine. (Stiftung Deutsches Technikmuseum Berlin/Historisches Archiv)

Oberleutnant Bethge, the commander of Jagdstaffel 30, achieved his 20th victory on 10.3. I would like to express my fullest recognition to the brave and proven flying officer and best wishes for further successes. (signed) von Quast.[1]

After Hans von Holthusen's Kette had gone through an inconclusive combat with five D.H.4s on the morning of March 16, 1918, machines from the Staffel took off for the front again in the early afternoon. This time they had more success, as Lt. Hans-Georg von der Marwitz shot down an English two-seater over enemy territory at 14:20. Since the downing of the enemy plane was observed from the German side, he later received recognition for it as his 4th victory.

On the next day four aircraft of the Staffel took off under the leadership of Oblt. Hans Bethge, concerning which the following report is available:

Around 11:30 the Staffel took off under the leadership and at the command of the Staffelführer

Left: The funeral service of Hans Bethge at the church in Phalempin.

Above: Lt. Kurt Katzenstein in the pilot's seat of his Pfalz D IIIa, and in front his two mechanics. The signal pistol on the wing identifies him as the Kette leader. (Stiftung Deutsches Technikmuseum Berlin/Historisches Archiv)

Oberleutnant Bethge for an offensive patrol. Oberleutnant Bethge flew the aircraft Pfalz D IIIa 5888/17. Southwest of Roulers the Staffel attacked an enemy formation of approximately ten two-seaters. Oberleutnant Bethge approached the aircraft he was attacking within a distance of about 200 meters. It was also noticed how he went down in a sharp left turn. Since aircraft accompanying him were occupied with the other Englishmen, the end result of the aerial combat taken up by Oberleutnant Bethge could not be followed any further.[2]

Oblt. Hans Bethge crashed fatally in the vicinity of Passchendaele. The two-seaters attacked were possibly D.H.4s of 57 Sqn. RFC.

The Staffel was deeply affected by the death of "Papa Bethge," as he was also called by his pilots. Otto Fuchs, who with Jagdstaffel 77 had just survived a crash between the lines and was still suffering from shock, found out about the death of his fatherly friend through a telephone call from Hans-Georg von der Marwitz, and was deeply shaken. Every pilot of Jagdstaffel 30 tried to get over the loss in his own way. For example, Hans Holthusen climbed into his Pfalz the next morning and flew alone aimlessly for about half an hour behind the front in order to clear his head.

But the war allowed little time for grief. Already on the morning of the next day he and other pilots of the Staffel had an inconclusive combat with ten S.E.5s and Sopwith Camels.

As the highest ranking officer of the Staffel, the "Offizier z.b.V." Oblt. Kurt Preissler took over acting leadership of the Staffel for the time being. In the air the Staffel was again led by Lt. von der Marwitz, who once more proved his leadership qualities in this difficult situation. He gathered the deeply depressed pilots around him; he spoke with each individual and tried to give them courage, therewith sustaining aerial activity. To the great relief of everyone, he was named Staffelführer by the Kogenluft on April 16, 1918.

Above: Phalempin airfield, photographed on May 3, 1918 by an aircraft crew from Flg. Abt. 18, Lt. Hans Holthusen's former unit. In the upper part of the photo are the Lille - Douai rail lines.

On March 21, 1918, the day on which the German spring offensive began, the Staffel was deployed in the sector of the German 17th Army and amongst other things had to go through aerial combats with four Sopwith Camels and eight D.H.4s.

In the following week the Staffel was deployed regularly, but there were only sporadic aerial combats. Thus the logbook of Hans Holthusen records from the 22nd till the 30th of March thirteen flights at the front without a single aerial combat.

On March 30, 1918, the Staffel unexpectedly increased in size: Oblt. Harald Auffarth, the Staffelführer of neighboring Jagdstaffel 29, landed at Phalempin "to lend support in the leadership of the Staffel," as is inappropriately stated in the transcript of the Jagdstaffel 30 War Diary by Erich Tornuss. Behind the event was a quite simple background which was apparently misunderstood by the military layman Erich Tornuss. Hans Bethge had not only been the Staffelführer of Jagdstaffel 30 but also the leader of Jagdgruppe "Nord" of the 6th Army. Harald Auffarth had now become his successor as commander of the Jagdgruppe. He could not, however, fulfill his new task from Bellincamps, as the necessary technical requirements (e.g., telephone connections) were not available there. As a result, he was forced to use the existing equipment at Jagdstaffel 30. As an initial measure, he saw to the transfer of his Jagdstaffel 29 to Phalempin, which was already moved by the next day.

From this time onward, both Staffeln frequently flew their missions together. According to the War Diary of Jagdstaffel 29, the Staffel remained at Phalempin till August and then moved again to Bellincamps. (Incidentally, in this war diary Jagdgruppe "Nord" of the 6th Army is designated as "Jagdgruppe 5," which indicates a renaming of the unit in the meantime.)

This month, which was so sad for the Staffel, ended with a further loss. Uffz. Max Marczinke, who had been with the Staffel for just two weeks, was fired upon above Ploegsteert Wood by British flak. He had to make an emergency landing in enemy territory with bad hits to his machine and fell into British captivity.

In Expectation of the New Fokker D VII

In the first week of April the Staffel had to carry out only occasional aerial combats. In this period the logbook of Hans Holthusen records only one aerial combat, with three Sopwith Camels on April 7, 1918 in the German 17th Army sector.

After aerial activity had been considerably hindered by fog and rain in the week of April 5 to 10, 1918, with the weather clearing up on April 11 and 12 it increased greatly, especially in the area of the German attack. According to their mission, the German fighters were to intercept the enemy at a low altitude in order to prevent attacks against German ground troops.

On April 11, 1918 Hans Holthusen already had an unsuccessful combat with an R.E.8 in the afternoon. On the next morning he and other Staffel comrades under the leadership of the acting Staffelführer got into an aerial combat with Sopwith Camels of 54 Sqdn. RAF around 8:30. In this fight Lt. Hans-Georg von der Marwitz succeeded in forcing his opponent to land in German territory. The wounded British pilot fell into captivity.

Around 11:00 the Staffel was again deployed in the area of the German attack and again ran into Sopwith Camels and some S.E.5s. While Hans Holthusen returned home from this fight without success, Lt. von der Marwitz succeeded in shooting down a Sopwith Camel near Aubers. This time the enemy pilot died in the crash. On the afternoon of the same day Hans Holthusen was again in action and had an inconclusive aerial combat with D.H.4s and S.E.5s near Armentières.

In the weekly aviation report of A.O.K. 6 for the week of April 12 to 18, 1918 it is stated regarding the deployment of the Jagdstaffeln of the German 6th Army:

Our Jagdstaffeln sought out the enemy at the lowest altitudes and successfully hindered him in the harassment of the troops. Their success in the course of the week – 24 victories – which were all achieved at low altitudes, prove that the fighter pilots are learning to adjust to their latest mission, freeing the troops of low-level enemy flyers.[3]

With rain and sleet and gusty winds on April 19, 1918 there again reigned so-called "flyers' weather," i.e., there could be no flying. So the pilots had enough of an opportunity to celebrate the promotion of Vzfw. Hans Oberländer to Leutnant der Reserve. At the same time, according to Hans Holthusen, the popular "z.b.V." Oblt. Preissler had to be bid farewell. He had been transferred to the staff of Jagdgeschwader II.

On the next day the weather improved and Hans Holthusen took off, with Oblt. Preissler in the observer's seat, at the controls of a Halberstadt for Flg.Abt. (A) 209 at Toulis near Marle. Then

Above: O In front of one of the airplane hangars. From left: Lt. Holthusen, Lt. von der Horst, Lt. von der Marwitz.

they continued to Jagdstaffel 63 at Balâtre near Roye. After a discussion there the flight path led finally to Jagdstaffel 15 at Ham, where the staff of Jagdgeschwader II resided. From there Hans Holthusen flew on a direct path back to Phalempin.

Four days later, on April 24, 1918, an old acquaintance, Lt. Kurt Katzenstein, arrived again at Phalempin. With his arrival the Staffel had one more experienced fighter pilot at its disposal, who in addition enjoyed great popularity within the circle of comrades. Lt. von der Marwitz immediately appointed him as the leader of a Kette. According to a statement by Hans Holthusen, in the period of April to June 1918 the Kette leaders were: Lt. von der Marwitz as Staffelführer (Kette 1), Lt. Holthusen (Kette 2), Lt. Oberländer (Kette 3), and Lt. Katzenstein (Kette 4).

In May 1918 there was a series of personnel changes in the Staffel. Lt. Rudolf Freiherr von der Horst was placed at the disposal of Idflieg. After eight months' service as a fighter pilot he was no longer up to the physical and psychological demands. On the other hand, Lt. Hans Eggersh, Uffz. Paul Marczinski, and the once more recovered Lt. Reinhold Maier arrived at the Staffel. As a

Above: The new commander of Jagdgruppe Nord, Oblt. Harald Auffahrt, at Phalempin airfield in May 1918. Sitting, from the left: Oblt. Auffahrt (commander of Jagdstaffel 29) and Lt. Hans Holthusen with his dog "Puck". Standing, from left: Lt. Kurt Katzenstein (Jagdstaffel 30) and Lt. Heinrich Nebelthau (Jagdstaffel 29).

successor to the transferred Oblt. Preissler, Lt. Ewald Siempelkamp became "Offizier z.b.V." According to a statement and documents of Erich Kaus, in the same month the Staffel received some Roland D VIa's for testing at the front, which, however, did not really reveal any improved flight performance in comparison with the Pfalz D IIIa.

In the middle of the month aerial activity, especially on the Allied side, increased considerably. The German Jagdstaffeln were therewith confronted by an ever more overwhelming number of Allied airmen, as the Kofl announces in the weekly aviation report:

The fighter pilots were mostly powerless against the strong British formations, and despite relentless deployment they did not always succeed in driving back the superior enemy forces and paving a clear path for our work planes.[4]

This increase in enemy activity is also reflected in the logbook of Hans Holthusen. While between May 2nd and 9th there is contact with the enemy in only five of a total of 14 frontline flights, in the eight days from May 11th to 19th there were combats with enemy flyers in 11 of 14 frontline flights.

On May 16, 1918 Pfalz D IIIa's from the Staffel under the leadership of Lt. von der Marwitz attacked a formation of Sopwith Camels of 208 Sqn. RAF. During the dogfight the Staffelführer succeeded in getting behind an opponent and forcing the airplane to land in German territory at 8:40. The British pilot, Lt. Cowan, climbed from his machine unwounded. On the next day he was the guest of Jagdstaffel 30, before being transported to a prisoner-of-war camp.

During a patrol at the front on the morning of the next day Hans Holthusen and his Kette first had a combat with five S.E.5s. Later the Staffel ran into ten Sopwith Camels. Lt. Holthusen succeeded in getting behind a Camel and forcing it down to an altitude of about 500 meters near Merville, on the other side of the front. But in the end his opponent succeeded in

Right: Lt. Hans-Georg von der Marwitz inspects Sopwith Camel D9540 of Lt. W. E. Cowan, which was forced to land by him during aerial combat. (Stiftung Deutsches Technikmuseum Berlin/ Historisches Archiv)

Above: Lt. Hans-Georg von der Marwitz take a close look to the Sopwith Camel of Lt. W. E. Cowan. (Stiftung Deutsches Technikmuseum Berlin Historisches Archiv)

Above: Hans-Georg von der Marwitz with his mechanic in front of the Sopwith Camel D9540. His mechanic holds the British Cockade, cut out as a souvenir. (Stiftung Deutsches Technikmuseum Berlin/ Historisches Archiv)

escaping in the ground mist.

Two days later Pfalz D IIIa's of Jagdstaffeln 29 and 30 took off for the front in a group formation under the leadership of Oblt. Auffarth. After an inconclusive skirmish with twelve Bristol Fighters, they attacked a formation of twelve S.E.5as a short time later. Lt. Eggersh succeeded in downing an opponent, but the success was awarded to Uffz. Pech of Jagdstaffel 29, who had likewise submitted a claim. In the same aerial combat Hans Holthusen was forced down by an S.E.5a to about 50 meters, but he was just barely able to shake off his opponent and land unscathed at Phalempin.

On May 20, 1918 the Ketten of Lt. Holthusen and Lt. Oberländer attacked three S.E.5as of 40 Sqn. RAF. Hans Holthusen, together with Hans Oberländer, succeeded in forcing down an opponent over Béthune. The English airplane finally crashed under the fire of the latter between Vieille Chapelle and Locon.

The next day the Kette of Hans Holthusen once more had an aerial combat with three S.E.5as. At 11:00, after a long dogfight, Hans Holthusen succeeded in forcing an opponent to land on territory recently conquered by German troops between Calonne and Pacaut Wood, about 4 km. south of Merville.

Above: Lt. von der Marwitz with his opponent of May 16, 1918, Lt. W. E. Cowan of 208 Sqn. RAF. Lt. Cowan was a guest at the Staffel for several days.

On May 23, 1918 Lt. Oberländer was badly wounded in the shoulder during an attack on an enemy formation. He was however able to make an emergency landing in his machine on German-held territory before losing consciousness.

Four days later, on May 27th, Vzfw. Arthur Schiebler, who had been with the Staffel for just five weeks, was shot down in his Pfalz D IIIa in aerial combat with S.E.5as at 21:00 and crashed fatally near Douvrin.

Jagdstaffel 30's opponents belonged to 40 Sqn. RAF and after their return wrote the following reports:

Combat Report of Maj. Roderic Stanley Dallas in S.E.5a D3520:

I dived on the closest enemy aircraft and fired 100 rounds from both machine-guns at close range. I followed the enemy aeroplane down and saw it crash in the vicinity of Hantay. I spotted another crashed enemy aircraft burning at the wood east of Billy.

Combat Report of Capt. Gwilym Hugh Lewis in S.E.5a D3540:

I fired a long burst from both machine-guns at a Pfalz with a white tail. The motor of the aeroplane seemed to stop in a cloud of smoke, then the aeroplane crashed, bursting into flames, near Hulluch.

Combat report of Lt. Ivan Frank Hind in S.E.5a B675:

I observed eight enemy aircraft, who were crossing over our lines at 14,000 feet. I went into a dive and got into good firing position behind a Pfalz with a silver fuselage. I fired about 100 rounds from close range, until I had to turn away to avoid a collision. I saw the enemy aircraft crash in the vicinity of the dynamite factory to the east of Billy.[5]

It remains to be noted that besides Vzfw. Schiebler no further German loss can be ascertained.

A further four days later, on May 31, 1918, Lt. Erich Kaus was wounded in aerial combat and admitted to the military hospital, about which he

Above: Crash landing by Lt. Reinhold Maier in May 1918. According to Hans Holthusen, this was not an unusual occurrence.

reported:

An English fighter came towards me, somewhat higher than I. Because one rarely obtains a hit during such maneuvers, I was surprised that a phosphorus round shattered on my left machine-gun. Then I felt blood running in my glove. A large part of the round had penetrated through my fur glove into my lower arm. Because I didn't know how severe the wound was and I feared losing consciousness through loss of blood, I let myself spin down and flew back to the Staffel at a low height. Fortunately only a tendon was shot through and splinters had to be removed in an operation.[6]

The loss of two experienced pilots like Lt. Oberländer and Lt. Kaus was another bitter setback for the Staffel. In addition, Lt. Holthusen went on a four-week furlough. As Erich Kaus later remarked, at this time the partly worn-out Pfalz D IIIa's of the Staffel were at best good enough for presenting targets for enemy aircraft. One could only achieve successes through surprise attacks. In a dogfight one was clearly at a disadvantage, except when one encountered a completely inexperienced opponent.[7]

How strong the numerical superiority of the Allied formations was in the meantime is also shown by the weekly aviation report of AOK 6 during the week of May 31 to June 6, 1918:

The front was flown over 234 times by formations which altogether numbered 1363 airplanes. The formations varied between 3 to 14 aircraft.[8]

In June 1918 the Staffel only achieved three victories, two of them on June 9, 1918. Two Ketten of Jagdstaffel 30 under the leadership of Lt. von der Marwitz and Lt. Katzenstein had run across Sopwith Camels of 210 Sqn. RAF. Lt. Hans-Georg von der Marwitz succeeded in forcing an opponent to land on German territory near Vieux Berquin at 9:11. Only nine minutes later Lt. Katzenstein was able to force down another Camel and likewise make it land. Both pilots fell into captivity, one of them badly wounded.

On June 17, 1918 Lt. von der Marwitz was badly wounded in the thigh, supposedly by friendly groundfire. Fortunately he succeeded in making a smooth emergency landing on the German side of the lines. For the time being, Lt. Eggersh took over acting leadership of the Staffel.

Till the end of the month, the only other success was achieved by Lt. Maier on June 21, 1918. He was able to send down an enemy captive balloon in

Above: The Pfalz D IIIa of Lt. Erich Kaus in May 1918. The aircraft is painted in the colors of the old Kingdom of Hannover, yellow and white. (Stiftung Deutsches Technikmuseum Berlin Historisches Archiv)

Below: Maintenance work on the same Pfalz D IIIa.

Above: Lt. Erich Kaus in the pilot's seat of Roland D VIa D.1219/18 in May 1918.

flames south of Béthune at Aix-Noulette.

On the next day the leader of Jagdgruppe 3, Oblt. Richard Flashar, was entrusted with the leadership of the Staffel, but this was only a nominal appointment. Oblt. Flashar only flew occasional patrols. In addition, the fighting strength of the Staffel was exhausted. For this reason a ten-day rest period was ordered for the Staffel. At this time Lt. Otto Franke and Lt. August Hartmann arrived at the Staffel. Otto Franke came from Brückstett near Langensalza and beforehand had served with Flg.Abt. 5 for nearly three weeks. August Hartmann was from Frankenthal. Before his training at Jagdstaffelschule I he had completed some missions with Flg.Abt. (A) 253.

Markings and Paint Schemes of the Aircraft

The majority of the airplanes bore the paint scheme already described in the previous chapter without any changes. Only the national insignia was changed in this period from the iron cross to the straight-armed "Balkenkreuz" like the Pflalz D IIIa of Hans Holthusen (Profile 47).

There were however some added touches to individual aircraft. Thus when Lt. von der Marwitz took over the leadership of the Staffel in April 1918, he also had the upper surfaces of his Pfalz D IIIa's wings painted wine red in order to be immediately recognizable as the Staffelführer (Profile 48).[9]

Kurt Katzenstein, who came back to the Staffel in April 1918 after an absence of three months, had the fuselage of his Pfalz D IIIa painted black. On the top surface of the upper wing was emblazened the orange-colored lozenge edged in black, which was supposedly customary for Kette leaders (Profile 49).

The Pfalu D IIIa of Erich Kaus had yellow and white lengthwise stripes and thereby represented again the color of the Kingdom of Hannover (Profile 50).[10]

It is not known whether the few Roland D VIa's sent to the Staffel for frontline evaluation had personal paint schemes or the Staffel marking.

Above: Hans-Georg von der Marwitz (second from left) in the circle of his pilots after the end of his flying service. One can recognize Lt. Hans Holthusen as the third from left.

Above: Pilots Erich Kaus, Hans Holthusen, and Hans-Georg von der Marwitz in the garden by their quarters. Naturally the dogs Puck and Duc are also there, as always.

Right: Lt. Hans Eggersch. (M. Szigeti)

Above: The Pfalz D IIIa of Lt. Holthusen in May 1918 with national insignia which have been altered in the meantime. From left: Uffz. Funk, Lt. Kaus, Lt. Mayer, Lt. Eggersh, Lt. Oberländer, Lt. Simpelkamp, Lt. von der Marwitz, Lt. König (guest) (Stiftung Deutsches Technikmuseum Berlin/Historisches Archiv)

Below: Pilots of the Jagdstaffel in front of the readiness hut May 1918. From left : Lt. Holthusen, Lt. von der Marwitz, Lt. Kaus, Lt. Oberländer, Lt. Mayer, Lt. Simpelkamp, Uffz. Funk and Lt. König (guest) (Stiftung Deutsches Technikmuseum Berlin/Historisches Archiv)

6. July 1918 – November 1918

Above: The wine red Fokker D VII of Lt. Hans-Georg von der Marwitz. The light blue painted front and underside of the engine cowlings, as well as the likewise light blue painted undercarriage, are clearly seen. (Stiftung Deutsches Technikmuseum Berlin/Historisches Archiv)

With the Fokker D VII Up Until the Armistice

In order to fill the post of Staffelführer, which had been vacant since the wounding of Hans-Georg von der Marwitz, on July 1, 1918 the Kogenluft ordered the appointment of Lt. Kurt Müller (Jagdstaffel 24) as commander of the Staffel. The take-over of the Staffel by the new Staffelführer, however, was delayed, since at the moment he was thoroughly needed at his old unit, and so Lt. Holthusen, freshly returned from leave, again took over acting leadership for the time being.

According to his own statement, Hans Holthusen could not remember "a Lt. Müller" at the Staffel, from which one can deduce that Kurt Müller did not officially take over the Staffel at all. Whether Lt. Müller in fact never was with the Staffel has not been passed on; if he was there, then it was probably for only a few days, during which he (logically) left behind no "lasting" impression.

The outstanding event of the day for the Staffel was something completely different. After operations had to be contested for a year with inferior airplanes, the first Fokker D VIIs could now be taken over. On July 4, 1918 Lt. Holthusen carried out his first operational flight with the new Fokker. While doing so the machine reached a maximum altitude of 5000 m. and a flight duration of 75 minutes.[1] The pilots were enthusiastic about the flight characteristics of their new machines. In a conversation with one of the authors, Hans Holthusen described the Fokker D VII as the best and most reliable airplane he had ever flown.

His logbook delivers eloquent proof of this reliabilty mentioned by him. While in the months from January to June 1918 he had to complete his 116 operational flights with seven different machines because of the high unreliability of the Pfalz D IIIa, from July until September 1918 (the time of his transfer to Jagdstaffel 29), he carried out a total of 82 operational flights on the Fokker. And he flew all of these on the Fokker D VII assigned to him, with serial number D.370/18.

With the Fokker D VII the German fighter pilots finally had an airplane once more which was equal,

Right: The last Staffelführer of Jagdstaffel 30, Lt. Hans-Georg von der Marwitz with is dog "Duc" in the garden of the crew quarter at Phalempin (Stiftung Deutsches Technikmuseum Berlin/Historisches Archiv)

if not superior, to the Allied fighter planes and gave them the possibility of at least offering perceptible resistance to their opponents attacking in ever larger formations.

On the afternoon of July 5, 1918 Hans Holthusen had his first (inconclusive) aerial combat at the controls of his Fokker D VII D.370/18, when he and his Kette attacked two D.H.9s. On the morning of the next day, under the leadership of Hans Holthusen, the Staffel had a combat with 8 S.E.5s and Sopwiths, which likewise remained without any significant result. After the frontline patrol on the morning of July 7th had proceeded uneventfully, in the afternoon of the same day there was again an aerial combat with five Sopwiths and two Bristol Fighters without result.

On July 8, 1918 Vzfw. Hermann Benzier, Vzfw. Ernst David, and Uffz. Ehrlich were transferred to the Staffel. All three were novices and came directly from the Jagdstaffelschule.

The patrol on the morning of this day first brought an aerial combat with five Sopwiths, and afterwards another one with five S.E.5s. Finally, the Staffel also attacked eight Bristol Fighters. All three aerial combats proceeded without result. In the afternoon several machines from the Staffel were involved in a combat with two Bristol Fighters and five Camels. This encounter too led to nothing of note.

On July 10, 1918 the Staffel with its new Fokker D VIIs, together with a Kette from Jagdstaffel 52, attacked seventeen Sopwiths. While doing so, Hans Holthusen was able to separate one of the British pilot's from his formation and after a long fighting pursuit to force him down to 20 meters altitude, and finally to make him land on the slopes of Mount Kemmel. The pilot was injured upon impact and taken prisoner. As Hans Holthusen later explained, in his estimation the Fokker D VII was superior to the Sopwith Camel. Lt. August Hartmann was

Above: Lt. August Hartmann's Fokker D VII with his personal marking of a witch on a broom.

supposedly lightly wounded in this air combat. However, he succeeded in bringing his airplane to a smooth landing.

On the morning of July 14, 1918 the Fokkers of the Staffel first attacked a D.H.4, which was, however, able to escape. Afterwards at 9:40 there was an aerial combat with approximately 20 S.E.5as over enemy territory near Merris. Lt. Hans Holthusen and Lt. Reinhold Maier each reported shooting down an opponent. However neither of these successes were confirmed. It seems possible that both downed aircraft were also claimed by two pilots from Jagdstaffel 40, which also participated in the combat, as they received confirmation for two S.E.5as.

The next day Lt. von der Marwitz returned from the hospital and again took over leadership of the Staffel. Even though he was limping, since the wound had not completely healed, he did not want to leave the Staffel in the lurch. This was, according to Hans Holthusen, in keeping with his sense of duty. This was taken into account by the Kommandierender General der Luftstreitkräfte, who appointed him again officially as Staffelführer on July 25th. The appointment of Lt. Kurt Müller was cancelled, who a few days later took over Jagdstaffel 72, which had been "orphaned" in the meantime.

On July 17, 1918 the Staffel, under the leadership of its old and new Staffelführers, attacked what Hans Holthusen described as a "mixed formation" of British aircraft. In the ensuing combat Lt. Otto Franke succeeded in inflicting several hits in the engine of a Sopwith Dolphin, whereupon the aircraft had to make an emergency landing in German territory near Erquinghem. The British pilot, who was not wounded, tried to set his airplane on fire, but this was prevented by German soldiers rushing up.

During this Lt. Franke was circling around the landing spot when a German machine-gun on the ground opened fire on the Fokker D VII by mistake and – which is "normal" in such cases – struck it fully. Lt. Franke had survived his first victory only by a few minutes. This incident was especially tragic in that Otto Francke had been the last survivor of four brothers in the German armed services. For this reason his mother had pointed out this circumstance especially in a letter to Hans-Georg von Marwitz, which led him to give his Kette leader, Hans Holthusen, the instruction "to pay special attention to the lad and, as far as possible, to keep him out of the biggest mix-ups."[2] However, nothing could be done about overly zealous machine-gunners behind one's own lines.

Several German fighter pilots of World War 1 with whom one of the authors spoke had bad things to say about the machine-gunners on the ground, as time and again they opened fire without certain

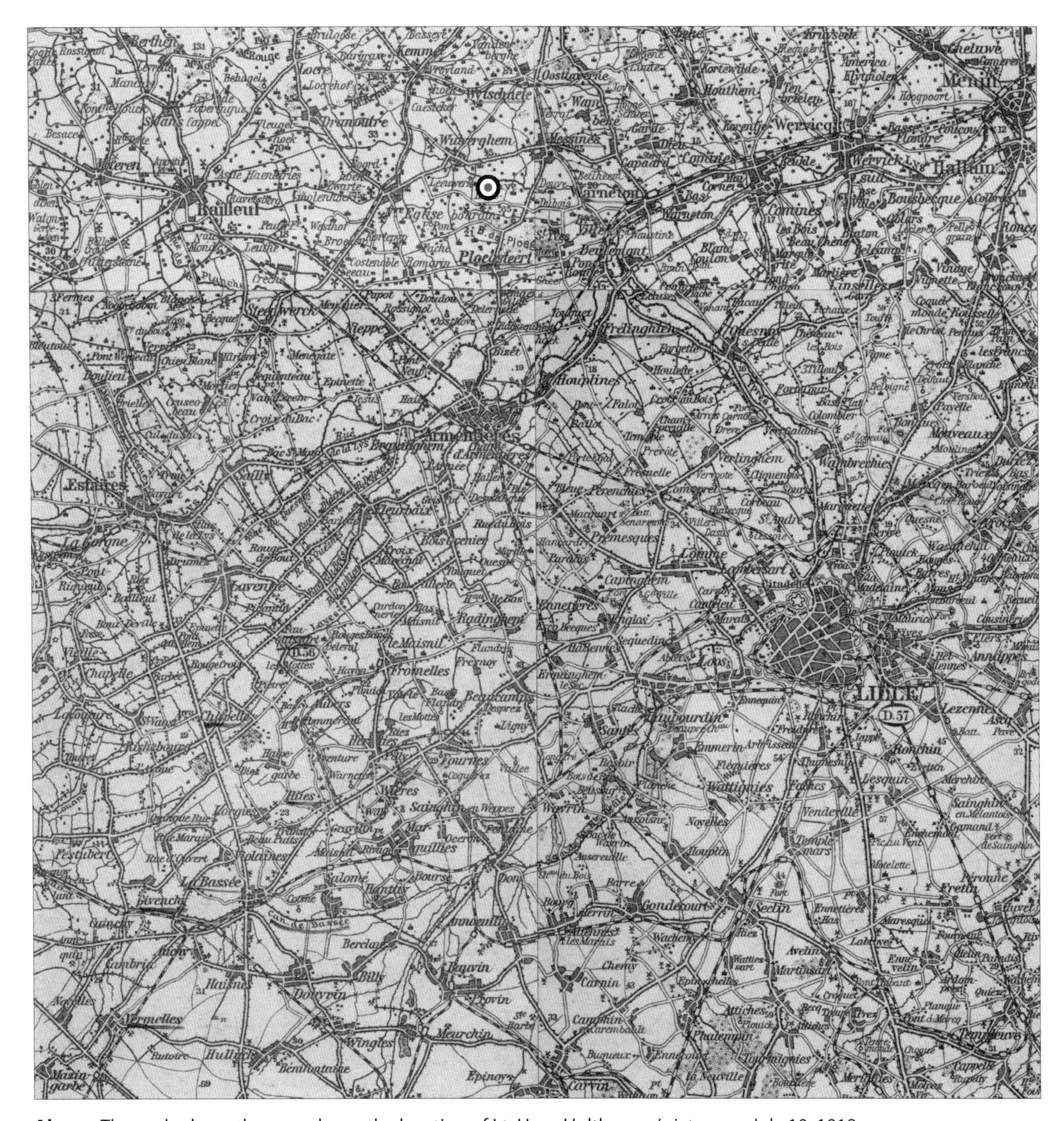

Above: The cockade on the map shows the location of Lt. Hans Holthusens' victory on July 10, 1918.

identification and so frequently fired upon their own aircraft. One reason for this, according to the opinion of veteran fighter pilots, especially towards the end of the war, was the increasing inexperience and poor training of the machine-gunners.

On the morning of July 19, 1918 the Staffel attacked ten D.H.4s. Lt. Holthusen, however, had to break off his attack due to a jam. He had the same experience on the morning of the next day during an unsuccessful attack on twelve D.H.4s.

Two days later the Staffel had a further loss to mourn. A Kette of three Fokker D VIIs from the Staffel was attacked by several S.E.5as of 40 Sqn. RAF at 9:50. In the process Uffz. Paul Marczinski was shot down near Pont Maudit and died. The following British report is available regarding this combat:

Above: A mechanic in the pilot's seat of the "Hexle" ("little witch"), as August Hartmann called his aircraft in his Palatine dialect.

Lt. Gilbert John Strange in S.E.5a D3527:

I attacked an enemy aircraft from a position of superior height from about 150 yards between Carvin and Pont-à-Vendin and fired 50 rounds at close range. Thereafter the enemy aircraft immediately spun down, turned over onto its back and slowly spun down out of control. I could not follow the enemy aircraft lower than 4000 feet... as the two other Fokkers were about to attack me.[3]

On the morning of July 24, 1918 a Kette from the Staffel under the leadership of Lt. Holthusen was again in action and attacked a D.H.9, but Hans Holthusen once again had to break off the attack due to a stoppage. On the evening of the same day Lt. Reinhold Maier succeeded in shooting down an enemy balloon in flames northwest of Grénay around 22:00.

Six days later the Staffel was aloft under the leadership of Lt. von der Marwitz in the morning in the northern sector of the German 6th Army. While on patrol they first ran into 10 to 20 airplanes described as "two-seaters," which were immediately attacked. After this attack had no success, Lt. von der Marwitz together with Hans Holthusen attacked a Sopwith Camel over enemy territory west of Merville, near the Nieppe Forest, which crashed at 12:30. It was the 9th victory of the Staffelführer.

Because his wound still was not completely healed, Hans-Georg von der Marwitz had to begin a two-week recuperative leave on the next day. Again Hans Holthusen, who had proven himself to be a capable Staffelführer, stood in as his replacement. In spite of that, the compulsory absence of a Staffelführer as experienced as Lt. von der Marwitz was, in view of the difficult position of the German fighter pilots, more than problematic for the Staffel. The weekly aviation report of the Kofl of the German 6th Army for the week of July 19 to 25 makes this clear:

The past week again brought new records in the number of enemy aerial "Geschwader" (formations) deployed. On July 22, 1918, the only day of uninterrupted flight activity made possible by the weather, the highest number of enemy formations ever reported in this Army on one day was attained

Vorlaeufiger Ausweis.

Im Namen

Seiner Majestaet des Kaisers und Koenigs

ist dem Gefreiten

Franz Moritz

am 5. August 1918

das Eiserne Kreuz II. Klasse

verliehen worden.

Leutnant und stellv.Staffelführer.

Felddruckerei IV. A.K.

Above: Interim document fort he bestowal of the EK II to Gefreiten Franz Moritz, Lt. Hans Holthusen signed on the 5th August 1918 as acting Staffelführer.

Above: Lt. August Hartmann carrying out target practice with a 98k carbine.

with 122 (enemy) formations. Fighter planes accounted for the greatest number by far; on the evening of the 22nd, after their flights continued unceasingly the entire day, in the area of Merville - Armentières - La Bassée approximately 50–60 enemy fighter planes could be counted at the same time.[4]

The term "Geschwader", it should be explained, would refer to any formation from a half-flight (3 machines) to a complete squadron (12 to 18 machines), or even more.

It remains to be added that Hans Holthusen had test-flown Fokker D VII D.541/18 on July 28, 1918. This airplane was one of three Fokker D VIIs which instead of metal fuselages had a fuselage made of wood. One wanted to be prepared in the event that a shortage of metal tubes should occur. Jagdstaffel 30 had received one of these prototypes for testing at the front.[5]

On the morning of August 1, 1918 Fokkers from the Staffel took off for the front under the leadership of Hans Holthusen and first encountered 8 British aircraft described as Sopwith Dolphins. However, the ensuing combat proceeded without results. On the afternoon of the same day Hans Holthusen with his Kette attacked a D.H.4. But while doing so they were inadvertently fired upon by German flak and had to break off the combat. Two days later in the morning there was an aerial combat with S.E.5s and Sopwith Camels over Meteren, which once more yielded no results.

For the 4th of August the logbook of Hans Holthusen records the ferrying of Pfalz D XII D.1488/18 from the Armeeflugpark to the Staffel.[6] This is possibly the same machine shown in a photo in the colors of Hans-Georg von der Marwitz after a crash landing. Unfortunately the date of this photo (taken in mid-August at the earliest) and the further circumstances of the crash landing are unknown.

The only thing certain is that the Staffel was at no time completely equipped with this competitor of the Fokker D VII. From this one must proceed with the assumption that at best one Kette flew the

Above: The crash-landed Pfalz D.XII in the colors of the Staffelführer, Lt. Hans-Georg von der Marwitz. Neither the date nor the person responsible for the crash landing are known. (Stiftung Deutsches Technikmuseum Berlin/Historisches Archiv)

Pfalz D.XII (which was the case in other Staffeln at this time). The reason for this actually undesireable "mixed operation" may be an insufficient number of available Fokker D VIIs for outfitting all Staffeln at the same time with a full complement of D VIIs. Consequently, one fell back on the new Pfalz D.XII in order to fill out the inventories. Incidentally, in spite of its strangely bad reputation, the Pfalz D.XII hardly took second place to the Fokker D VII in its flight performances.

Besides the Fokker D VIIs and Pfalz D.XIIs there were possibly still some Roland D VIa's in the Staffel inventory in August 1918. If however they were used at all, it would presumably have been for new pilots making test flights.

On August 6, 1918 Jagdstaffeln 14, 29, 30, and 52 were combined to form Jagdgruppe "Süd" (South) of the German 6th Army under the leadership of Oblt. Harald Auffarth.

The next day the Staffel took off under the leadership of the acting Staffel commander in a northerly direction and then turned towards the front in the direction of Armentières. About 5 km southwest of Armentières they ran into a British formation of 18 S.E.5s and Sopwith Dolphins. Vzfw. Ernst David was shot down in this aerial combat and crashed over Fleurbaix. (7) On the same day Vzfw. Hermann Benzier was injured during an emergency landing after an aerial combat. More exact details are unknown, but it may be assumed that this incident must also be attributable to the above-mentioned aerial combat.

On the morning of August 8, 1918 Hans Holthusen, in his capacity as acting Staffelführer, flew to a meeting at Jagdstaffel 34 in Ennemain and from there went to another meeting at Jagdstaffel 5 in Moislains. During this flight he was mistakenly fired upon by German flak. The reason for the meeting may have been the planning of operations of the Jagdstaffeln of the German 6th Army with the German 2nd Army on August 9 and 10, 1918. These operations were supposedly without results; in any case, nothing to the contrary is known.

As of August 11th Jagdstaffel 30 was again deployed in the sector of the German 6th Army. Up until August 14, 1918 the logbook of Lt. Holthusen records seven frontline flights, all of which however involved no contact with the enemy. The reason for this is found in the weekly aviation report of AOK 6 from the 9th to the 15th of August 1918:

In spite of the generally favorable weather, during the week being reported enemy aerial activity, after it had been relatively lively, has continuously decreased in an unmistakable connection to the clashes on the Amiens front.[8]

On August 15th Lt. von der Marwitz returned to the Staffel and, according to Hans Holthusen, took over the Staffel leadership as ever full of energy and

Fédération Aéronautique Internationale (F.A.I.)

Deutschland No 2303

Die unterzeichnete von der F.A.I. für Deutschland anerkannte Sportbehörde ernennt hiermit | Nous soussignés pouvoir sportif reconnu par la F.A.I. pour l'Allemagne certifions, que

Herrn Philipp Steiner

geb. zu Worms, den 11. Nov. 1896

nach Erfüllung aller von der F.A.I. vorgeschriebenen Prüfungsbestimmungen zum | ayant rempli toutes les conditions imposées par la F.A.I., a été breveté

Flugzeugführer | Pilote aviateur

für Zweidecker | pour Biplan

den 26. Mai 1917

Deutscher Luftfahrer Verband

Der Vorstand

Hergesell

Unterschrift des Inhabers.

Philipp Steiner

Above: The flight certificate, with number 2303, of Flg. Philipp Steiner, issued on May 26, 1917. (Kalin family)

thirst for action. Four days later, on August 19, 1918 around 8:15, about 40 British airplanes appeared over Phalempin airfield and began to attack it with bombs and machine-gun fire. Fortunately, the Staffel had taken off beforehand so that the attack did not cause any significant damage.

In spite of that, the Staffel was moved on the very same day to Avelin near Seclin (about 5 km northeast of Phalempin). The Staffel's operations were however hardly influenced by the move. Thus Hans Holthusen's logbook reports the next operational patrol on August 22, 1918. Between August 23 and September 5, 1918 the Staffel (like other Staffeln of the 6th Army) was frequently deployed in the sector of the neighboring German 17th Army to the south. This is established for the 25th, 27th, and 29th, as well as the 3rd and 4th of September.

In the process, on August 27, 1918 Lt. von der Marwitz reported the downing of a British aircraft described as a Sopwith in the British lines near Tilloy at 7:05. Two days later he succeeded in shooting down an S.E.5a in flames over German territory near Haplincourt at 8:10. Finally, on August 30th the news arrived at the Staffel that Hans-Georg von der Marwitz had been promoted to Oberleutnant (1st Lieutenant).

On September 1, 1918 Lt. Reinhold Maier took over the duties of Offz. z.b.V. from Lt. Simpelkamp, who had been sent to the hospital due to an old injury. Lt. Maier was the only Offz. z.b.V. who in addition to his regular duties also flew frontline patrols with the Staffel. As Hans Holthusen recalled, he was an enthusiastic pilot who would very much rather sit in the cockpit of his airplane than in the office.

In the diary of Josef Raesch (in 1918 a pilot in Jagdstaffel 43, which was likewise deployed in the sector of the German 6th Army), excerpts of which were published in the American periodical the *Cross & Cockade Journal*, Jagdstaffel 30 is also mentioned. Thus it is stated there under the date August 28, 1918:

...drive to the theater in Lille. There during intermission we meet Lt. von der Marwitz, commander of Jagdstaffel 30... Von der Marwitz likewise has no machines in his Staffel [like Jagdstaffel 43 at this time – authors] and cannot take off. He is very depressed and also worried...[9]

This note is astonishing, because in the time period concerned (see above) operations by the Staffel are established (even with victories). So at least "some" machines were fit for operations. In addition, the entry is not at all present in the original diary (to which the authors have access). From this one can only conclude that Josef Raesch at that time brought to publication a version of the diary revised by him, in which he expanded or supplemented passages from his memory.

With respect to Jagdstaffel 30 being "put out of action," it would appear that he erred in his memory (which is not surprising in view of the events having occurred almost 50 years previously at that time), because in the weekly aviation report of Kofl 6 for the week of August 16 to 22, 1918 it states in connection to the bombing raids on German airfields, etc. which took place in the army sector between the 16th and 19th of August: *"...JAGDSTAFFELN 43 AND 63 WERE PUT COMPLETELY OUT OF ACTION."*[10] There are in this connection no aircraft losses mentioned for Jagdstaffel 30.

A further entry in Josef Raesch's diary is likewise of interest, as here too it directly concerned Jasta 30.

5 September 1918. A regulation has come out that each Staffel will receive only 2000 liters for each ten-day period. If one calculates 10 aircraft like we have with 100 liters consumption after every flight, then we will run dry after two days. So things have to be divided up very neatly and only a certain number of aircraft may fly.[11]

The word "regulation" refers to secret order No. 69726/1016, Kofl 6 Ia/Ic, Army H.Q. of September 3, 1918, issued by Kofl 6. This reads as shown in the text box at the bottom of the page.

In the weekly aviation report of the 6th Army Kofl for the week of September 6 to 12, 1918 the results of the aviation fuel rationing were explained as follows:

Taking into account the prevailing fuel shortage, only two Jagdstaffeln were deployed alternately according to the combat situation in the air. The other Staffeln remained at rest, but were prepared for any required action.[13]

With this rationing, the deployment possibilities for the German Jagdstaffeln in the sector of the German 6th Army were considerably restricted.

It is conspicuous that in the period of August 28 to September 15, 1918 there are no flights recorded in Hans Holthusen's logbook. The cause of this is unknown; one could perhaps trace it back to the above note from Josef Raesch's diary, but it is more likely due to a period of home leave.

In the most extremely difficult situation in which the German Jagdstaffeln found themselves, an old acquaintance returned to Jagdstaffel 30 on September 17, 1918 in the person of Lt. Hans Oberländer, who with his experience represented a welcome reinforcement.

On the same day 21-year-old Flieger Philipp Steiner arrived at the Staffel. He came directly from Jagdstaffelschule I, but had a long flying career behind him. Already in October 1915, after his training as a pilot, he was transferred to Feld-Flieger-Abteilung 23. A half year later he became a flight instructor at FEA 6 in Grossenhain and later at FEA 9 in Darmstadt. He had both several years experience as a pilot as well as frontline experience, which qualities also made him a valuable addition to the unit's flying personnel.

In accordance with Kogenluft 275530 Fl.III the consumption of aviation fuel by the flying units will as of September 1, 1918 be rationed on a ten-day basis with consideration of the battle situation. This rate of consumption allowed by the Kogenluft over a ten-day period falls considerably short of the average corresponding consumption by the flying units in the preceding months:

Unit	Avg. use during 10 days in July	Avg.daily amount of fuel in July	Avg. daily flight hours in July	For first 10-day day period in September is allowed	Avg. daily fuel amount 1st 10-days September	Avg. daily flight hours 1st 10-days September
Jasta	5670 liters	567 liters	11 hours	3200 liters	320 liters	6½ hours

An increase in the rate is not anticipated in the foreseeable future.[12]

On September 22nd a Kette from the Staffel consisting of Lt. von der Marwitz, Lt. Holthusen, and Lt. Bieling had an aerial combat with three Sopwith Camels over enemy territory near Ploegsteert Wood, in the course of which the Staffelführerachieved his 12th victory. Lt. Bieling was able to shoot down a further Sopwith Camel, whose crash in the wood was observed from the German side. In the afternoon of the following day Lt. Holthusen was on patrol with his Kette, without encountering enemy aircraft. On September 24, 1918 the same three Fokker D VIIs had an inconclusive combat with seven S.E.5as.

Hans Holthusen undertook his final frontline flight with the Staffel on September 27, 1918. Two days later he was transferred to Jagdstaffel 29 as acting commander. His departure from Jagdstaffel 30, above all from his old comrades, was very difficult for him too. Some decades later in a conversation with one of the authors he was still full of praise for Hans-Georg von der Marwitz, who according to Holthusen had developed from a carefree cavalry Leutnant to a responsible and charismatic Staffelführer.

One day before the transfer of Hans Holthusen Jagdstaffeln 30. 43, 52, and 63 were combined to form the Jagdgruppe of the German 6th Army under the leadership of Oblt. Adolf Gutknecht (Staffelführer of Jagdstaffel 43). On September 30, 1918 the Staffel moved to the airfield at Sin, only to move yet again only two days later to the airfield at Baisieux on the road from Lille to Tournai.

At this time Jagdstaffel 30 participated in the battles between the Canal de la Deûle and the Schelde River and the battles around the Hermann Line and the fighting withdrawal before the Antwerp-Meuse Line.

On October 8, 1918 Hans-Georg von der Marwitz succeeded in shooting down an S.E.5a as his 13th victory. Two days later the Staffel moved to Awaing airfield on the road from Leuze to Renaix. It was while operating from this field that the Staffel mourned its final losses. Lt. Reinhold Maier took off for a practice flight and "wiped out" in the immediate vicinity of the airfield from a low height while in a turn. He was rescued with most severe head injuries, but died while being transported to the closest military hospital.

On October 21, 1918 the Jagdgruppe of the German 6th Army was reinforced by Jagdstaffeln 28 and 33 and renamed Jagdgruppe 7b. "Pour-le-Mérite" recipient Lt. Emil Thuy (Staffelführer of Jagdstaffel 28) became the new Jagdgruppenführer. In view of the hopeless military situation, this organizational measure could no longer have any sort of influence on the further course of the war.

Above: Philipp Steiner in June 1918 as a flight instructor at Flieger-Ersatz-Abteilung 9 in Darmstadt. (Kalin family)

On October 27, 1918 the Staffel's Fokker D VII's took off for an operational flight, in the course of which there was an aerial combat with English single-seat fighters. During the aerial combat the motor of Flg. Phillip Steiner's Fokker D VII was shot up by the machine-gun fire of one of the attacking English aircraft. Leaving a white trail of vapor behind, the Fokker tipped and went down in a steep glide.

The English pilot pursued the Fokker D VII, continuously firing, until he in turn was attacked by the Staffelführer Hans-Georg von der Marwitz, who drove him off and then shot him down. Phillip

Steiner succeeded in landing his Fokker D VII with a dead engine. However, he wrecked it while doing so. Since he did not know which side of the front he was on, he jumped out of his stranded airplane and hid in a nearby hedge, until he recognized the approaching soldiers as German by their helmets. The airplane was recovered and brought back to the Staffel for repairs. Philip Steiner took a part of the splintered propeller as a souvenir. It hung for years on the wall of his living room.[14]

This claim was Oblt. von der Marwitz last victory of the Staffel and his own 14th (according to the daily report of Kofl 6).

Any further success was not possible in view of the ever graver shortage of fuel and the mainly inexperienced pilots. But thanks to the judicious manner in which Oblt. Hans-Georg von der Marwitz commanded his Staffel, all of his young pilots came through unharmed. Also the other "old hands" of the Staffel – Lt. Hans Holthusen, Lt. Hans Oberländer, and Lt. Kurt Katzenstein – had their own share in this.

The final loss of the Staffel occurred on October 28, 1918 when Vzfw. Alfred Jaeschke had to make an emergency landing on enemy territory near Tournai and fell into British captivity. The last move of the Staffel caused by the constant withdrawal of the front occurred on October 31st and brought them to the airfield at Vollezeele.

Finally, shortly before the end of the war Oblt. Hans-Georg von der Marwitz was awarded the Knight's Cross of the House Order of Hohenzollern in recognition of his achievements.

When the war ended with the armistice of November 11, 1918, Jagdstaffel 30 had achieved in its barely two years of existence a total of 64 aerial victories. The most successful pilots of the Staffel were Oblt. Hans Bethge with 17 (total 20), Oblt. Hans-Georg von der Marwitz with 13 (14), Lt. Hans Oberländer with 5 (6), and Vzfw. Josef Heiligers with 4 aerial victories. In the same period of time eight pilots of the Staffel had lost their lives in aerial combat, six more died in airplane crashes without enemy involvement, five were taken prisoner, and seven had been wounded. What followed thereafter can be quickly told: the tents at Vollezeele were taken down and the Staffel marched with its entire inventory back home, where still in November 1918 it was officially disbanded at a FEA (flyer replacement unit)/Flying School as ordered by the Kogenluft.

Markings and Paint Schemes of the Aircraft

Documentation regarding the painting of the machines in the Fokker D VII era is more than scanty. There exists a photo of the Fokker D VII of Hans-Georg von der Marwitz, two photos of the machine of August Hartmann, as well as the photo of a crash-landed Pfalz D.XII in the colors of Hans-Georg von der Marwitz. As a memento, Hans Holthusen had taken a photo of every pilot and his airplane before his transfer to Jagdstaffel 29, but these photos were irretrievably lost in his house in Hamburg in a firestorm in July 1943. Fortunately, with his help at least a few paint schemes could be reconstructed.

When the Fokker D VIIs arrived at the Staffel beginning on July 1, 1918 Lt. von der Marwitz was in the hospital, the designated new Staffelführer had not arrived yet, and Hans Holthusen was leading the Staffel in the air. For this reason the decision about painting the new aircraft lay with him. He took over the orange color of the lozenge.

Orange had been the Staffel color under Bethge and for me it was a matter of course that this tradition would be continued in remembrance of him, he explained regarding his decision.[15]

As a Staffel marking, the middle portion of the aircraft's fuselage was painted orange, while the colors of the nose and tail could be freely chosen by each pilot. Hans Holthusen commented regarding this:

With the orange-colored fuselage the airplanes were easily recognizable even in the thickest aerial combat, and I always succeeded in quickly assembling the Staffel again. My own Fokker had an orange-colored fuselage with a black motor cowling and a red-and-white horizontal tail (Profile 51).[16]

After Lt. Hans-Georg von der Marwitz had taken over the Staffel again, he did not change the Staffel color introduced by Hans Holthusen, but as a single exception he had the fuselage of his Fokker D VII—as usual—painted wine red. The side motor cowling was orange and the underside and front part of the cowling in his traditional light blue (Profile 52). According to Hans Holthusen, he later had a Fokker D VII which was painted identically, but bore the family crest on the fuselage The upper surface of the upper wing was painted orange so that his pilots could better recognize him in aerial combat (Profile 53).[17]

The Fokker D VII of Lt. Hartmann, who arrived at the Staffel at the end of June, had the orange-colored fuselage, and the motor cowling and the stern of the airplane were dark blue. On both sides of the fuselage was emblazoned a "witch on a broom." In his Palatine dialect August Hartmann always spoke of his machine only as "moi Hexle" ["my little witch"] (Profile 54).[18]

Epilog Part 1 – Personal Comments Regarding the Origin of this Book

There is a special fascination involved in writing a history based on objective facts and in part on documents available in only fragmentary form. However, this history becomes much more lively when one has a chance to speak with the participants in those past events and to incorporate their personal memories into such a history.

In 1975 an employee of the Deutsches Museum handed me the membership roster of the "Alte Adler" (Old Eagles), the association of pre-war flyers and pilots of the First World War. This membership roster opened for me the possibility of establishing contact with a number of former pilots of the German Jagdstaffeln, interviewing them, and while doing so having a look at photos and documents – to the extent that they were still available – and being able to copy them.

One of my first contacts was Otto Fuchs, a former pilot of Jagdstaffeln 30, 35, and 77. What was originally planned as a visit turned into a friendly relationship, and I had the joy of being able to be a guest of Otto Fuchs and Emilie Fuchs-Hussong again and again over the course of several years in Dachau, near Munich. During my visits I also learned the story of *Wir Flieger*, the book written by Otto Fuchs which was published in 1933.[1]

During the period of his membership in Jagdstaffel 30, Otto Fuchs had decided to publish his experiences as a pilot in the First World War in novel form after the war. For this reason, during his time as a pilot at the front he regularly took down notes of his experiences and missions; he recorded detailed notes from memory concerning conversations, and also took note of the paint schemes of his comrades' aircraft.

In 1932 he submitted his manuscript of about 500 pages to a publisher. The manuscript comprised three parts. Part 1 concerned his time with Flieger-Abteilung(A) 292, the second part dealt with the period of his service with Jagdstaffel 30, and the third part Jagdstaffeln 35 and 77. He had anonymized the names of his comrades, and the events in one of the three parts were often not presented chronologically. Also, the periods in which members of the Staffel appear in the novel did not always correspond to the actual chronology of events.

When the final book was sent to Otto Fuchs before printing, he was horrified. His original manuscript had been shortened by about half and "revised" by an editor. Important passages were missing from the second part, and the third part was rewritten as though the events concerned Jagdstaffel 30, although the persons described there belonged to Jagdstaffel 77.

The health problems resulting from his crash between the lines were left out. Passages which struck the editor as "unpatriotic" (i.e., not conforming to National Socialist language conventions) were deleted, and the person representing the Jewish flyer in the Staffel, Lt. Kurt Katzenstein (called Lt. Karl Goldstein in the original manuscript) no longer appeared. Also missing was the chapter "The Final Conversation," in which the Staffel commander, Oblt. Werner (Hans Bethge) in his disillusionment declares the war as lost.

Otto Fuchs tried to prevent the publication of the marred book, but only managed to have himself appear not as the author, but as the editor for an "anonymous friend."

When I gave Otto Fuchs my copy of his book with the request for a dedication, he wrote:

"It is not easy for me to write a dedication in a book from which I already distanced myself when it appeared. I portrayed myself as editor and not as author. The editor had cut it down to less than half of my original manuscript. This happened without my knowledge. In order to reestablish the disrupted connections, the scenes of events were combined and the group of people simplified... Persons emerged in which the characteristics and fates of several were combined. None of them entirely corresponds to his model. Genuine and unaltered are the experiences and the course of certain combats.

"If these are capable of conveying a picture of the lure and suspense of an aerial combat and the beauty of flying, then I am satisfied...

"With this in mind, dear Herr Schmäling, I place this book back in your hands."

Despite all abridgements and changes, Otto Fuchs' book is still a quite authentic account of the experiences of a German flyer in the First World War.[2]

With the kind assistance of the city archive of Hamburg I was able to get into contact with Hans Holthusen in 1979. My visit with Hans Holthusen was also an impressive experience. Unfortunately, Herr Holthusen had to inform me that all of his photo albums had been lost in a firestorm during the bombing raids on Hamburg in 1943. Besides his many memories, he still possessed his flight logbook, which thankfully he placed at my disposal. Thanks

Above: Otto Fuchs in conversation with Bruno Schmäling in the autumn of 1977 in the dining room of his house in Dachau.

to his kind assistance, the question of the Staffel marking used on the Fokker D VIIs, the subject of many a speculative discussion, could be answered conclusively.

Scottish air historian Alex Imrie had interviewed Erich Kaus in the 1960s, and placed the notes from these discussions at my disposal. He also helped me to establish a personal contact with Herr Erich Kaus, who was able to contribute further interesting details from his memory.

Already in the year 1958 Dr. Gustav Bock had visited and interviewed Hans Oberländer in Stuttgart. These notes as well have been incorporated into the book. Hans Oberländer also allowed him to take a glimpse at his photo album; unfortunately at that time there was no chance to make reproductions of the photos. (A German collector who in the meantime had come into possession of copies of these photos unfortunately refused to make them available for this work.)

In the 1960s Erich Tornuss was in contact with Erich Hartmann through letters, and received a few photos from this gentleman. Some contents of their correspondence, as well as the photos, were able to be included in this book.

Without the kind assistance of the former pilots of Jagdstaffel 30 mentioned above, it would not have been at all possible to point out many of the details

Above: Information from the British side at the *Cross & Cockade* meeting in London, August 1978. Bruno Schmäling is talking with Wing Commander G. H. Lewis DFC, 40 Squadron RAF, who took part in the air fight on the 27th May 1917 against aircraft of Jagdstaffel 30. Drinking together a pint of bitter, Wing Commander Lewis told Bruno Schmäling: "Your German pilots were damned hard fighting guys; I'm happy that we are now on the same side."

mentioned in this book. To them the authors owe their deep gratitude.

Bruno Schmäling

(1) Fuchs, Otto. *Wir Flieger: Kriegserlebnisse eines Unbekannten*. Leipzig: Koehler, 1933.

(2) In the meantime a very good English edition of *Wir Flieger*, with translation and commentary by Adam M. Wait, has appeared under the title *Flying Fox* (Schiffer Publishing Ltd., USA, ISBN 978-0-7643-4252-3). The German edition, on the other hand, can only be obtained through second-hand book dealers.

Epilog Part 2 – Members of Jagdstaffel 30 After the War

Only of limited information of the further life of members of Jagdstaffel 30 could be found.

Hans-Georg von der Marwitz stayed with aviation after the war and became a professional civil pilot. On the 12th May 1925 he fatally crashed during a test flight near ground level with the Mark-Eindecker while chief-pilot of the Stahlwerk Mark Flugzeugbau. In the crash he broke his neck.

Joachim von Bertrab also became a professional civil pilot after his return from captivity. He was killed in a crash on the 28th July 1922 near the village of Boitzenburg while pilot of the postal-airplane on the Berlin – Hamburg route together with three American passengers.

Otto Fuchs studied philosophy and literature from 1919 to 1920. In 1924 he studied aviation on the

Technical High School at Darmstadt and became a member of the "Akademischer Fliegergruppe" (academic aviators group). He interrupted his study from 1927 to 1930 to work as a technical director at the secret aviation test site at Lipezk (Russia) were the German air force tested aircraft together with the Soviets. From 1933 on he worked for the DVL "Deutsche Versuchsanstalt für Luftfahrt" (German Research Institute for Aviation) and at the end of the war he was a department manager. In this position he successfully defended the takeover of his department by the National-Socialist Organisation. After the war it was mainly due to his efforts that the DVL was not disbanded and later he became one of the presidents of the DVL. He was also a well-known landscape painter and active member of the artist society. He lived in the city of Dachau near Munich until his death in 1987.

Hans Holthusen became a successful businessman and lived in Hamburg until his death in the 1980s.

Kurt Katzenstein became a very well-known sport pilot. Together with the former fighter pilot Antonius Raab he founded the "Raab-Katzenstein Aviation Company". The business of the company was very successful, but with the world economic crisis the company went bankrupt in 1930. After the Nazi-party took power in Germany Kurt Katzenstein left Germany and immigrated to South Africa were he lived under the name Kurt Kaye.

Paul Erbguth studied engineering at the University at Chemnitz and until the Second World War he worked as an engineer for the Siemens Company. In the Second World War he had the rank of major and was in command of an aircraft repair and maintenance company. His son was also a military pilot in World War 2, flying transport aircraft. After the war he worked again for the Siemens Company as "Oberingenieur (chief engineer) at Hof and Nuremburg. After his retirement he moved to Hof were he died in 1973.

Erich Kaus joined the Luftwaffe in 1935. In 1941 and 1942 he was an Oberstleutnant and commander of the Fliegerschule/ Fliegerausbildungsregiment 31 (pilot school/pilots regiment) in Posen. From June 1942 he was commander of pilot school 126 at Gotha. After the Second World War he lived in Bad Neuenahr until his death.

Below: Kurt Katzenstein with his partners of his aircraft company.

Above: Signed photo of Hans Holthusen.

Above: A portion of the former airfield at Phalempin in 2009, looking in the direction of the Rue Jean-Baptiste Lebas. At one time there were two tent hangars in front of the hedge at right. (A. Wait)

Above: Former airfield at Phalempin in 2009. During the First World War two aircraft hangars built in L-form stood in front of the small hedgerow on the left and the large hedgerow. (A. Wait)

Right: Former airfield at Phalempin in 2009. On this spot stood the hangar with the captured rudders on the roof. In the background can be seen the steeple of the church where the funeral service for Hans Bethge took place. (A. Wait)

Above: Otto Fuchs in 1928 with the BAG (Bahnbedarf AG, Darmstadt) D.IIa, Werknummer 14, with the registration number D-893. The airplane was painted "bilious green" (emerald green), like his Albatros D III at Jagdstaffel 30. He added the fox of his Albatros D V, but in this case the fox was white.

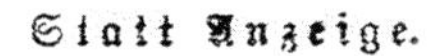

Statt Anzeige.

Am 28. Juli verunglückte tötlich durch Absturz mit dem Flugzeug unser einziger, inniggeliebter Sohn, unser lieber Bruder und Schwager 912 19

Joachim-Lambert von Bertrab,

Kgl. Oberleutnant a. D., früher im Niedersächs. Feld-Art.-Regt. Nr. 46 u. Jagdflieger. Ritter des E. K. I und anderer Kriegsauszeichnungen.

Neuroofen bei Menz i. Mark.

In tiefem Schmerz

v. Bertrab, Forstmeister.
Helene v. Bertrab, geb. Wichmann.
Ilse v. Buch, geb. v. Bertrab.
Georg v. Buch, Forstassessor.

Above: Obituary notice of Joachim-Lambert von Betrab. (O. von Betrab)

Right: Hans-Georg von der Marwitz after the war, ca. 1924. (A. Imrie)

Above: Lt. Erich Kaus with his two mechanics in front of Pfalz D IIIa 5981/17 Spring 1918 (A. Imrie)

Above: Albatros D III D.767/17 flown by Lt. Oskar Seitz. (Aaron Weaver)

Roster of Flying Personnel

1. Staffelführer

Rank	Last Name	First Name	Period of Service	Remarks
Oblt.	Bethge	Hans	15.1.17–17.03.18	KIA
Oblt.	Preissler	Kurt	19.03.18–16.04.18	JG II
Oblt.	von der Marwitz	Hans-Georg	16.04.18–11.11.18	

2. Acting Staffelführer

Rank	Last Name	First Name	Period of Service	Remarks
Lt.	Erbguth	Paul	20.04.17–20.05.17	Jasta 30
Oblt.	Preissler	Kurt	10.11.17–10.12.17	Jasta 30
Oblt.	Preissler	Kurt	15.01.18–09.02.18	Jasta 30
Lt.	von der Marwitz	Hans-Georg	17.03.18–19.03.18	Jasta 30
Lt.	von der Marwitz	Hans-Georg	20.03.18–01.04.18	Jasta 30
Lt.	Eggersh	Hans	17.06.18–22.06.18	Jasta 30
Oblt.	Flashar	Richard	22.06.18–31.07.18	Gruja 3
Lt.	Holthusen	Hans	01.07.17–15.07.18	Jasta 30
Lt.	Holthusen	Hans	01.08.18–15.08.18	Jasta 30

3. Pilots

Rank	Last Name	First Name	Period of Service	Remarks
Lt.	Bastgen	Wendel	26.01.18–16.02.18	POW
Vzfw.	Benzier	Hermann	08.07.18–07.08.18	accident
Lt.	von Bertrab	Joachim	06.03.17–12.08.17	POW
Uffz.	Beyer	Bruno	18.01.17–09.07.17	Afla 235
Flg.	Bieler	Erich	16.08.18–11.11.18	
Lt.	Bieling	Friedrich	24.08.18–11.11.18	
Lt.	Brügman	Heinrich	30.03.17–15.08.17	KIA
Vzfw.	Bucher	Franz	01.05.17–08.06.17	KIA
Oblt.	Buddecke	Hans-Joachim	15.02.18–08.03.18	Jasta 18
Uffz.	Busch	Otto	24.09.17–08.12.17	Aflup 6
			21.12.17–15.02.18	Aflup 6
Vzfw.	David	Ernst	08.07.18–07.08.18	KIA
Lt.	Eggersh	Hans-Herbert	14.05.18–11.10.18	Bogohl 4
Uffz.	Ehrlich		08.07.18–11.11.18	
Vzfw.	Eimbeck	Gustav	18.01.17–12.04.17	Aflup 6
Ltn.	Erbguth	Paul	10.03.17–28.12.17	Jasta 54
Uffz.	Foege	Wilhelm	17.10.17–22.12.17	KIA
Lt.	Forstmann	Hans	16.06.17–02.07.17	KIA
Lt.	Franke	Otto	30.06.18–17.07.18	KIA
Lt.	Fuchs	Otto	05.06.17–08.11.17	Bayer. FEA 1
Uffz.	Funk	Josef	20.05.17–28.06.18	FRA 10
Lt.	Hartmann	August	30.06.18–11.11.18	
Lt.	von der Horst zu Hollwinkel	Rudolf	22.08.17–06.05.18	Idflieg
Vfw.	Heiligers	Josef	20.01.17–20.11.17	killed in crash
Lt.	Holthusen	Hans	06.10.17–29.09.18	Jasta 29
Vfw.	Jaeschke	Alfred	19.08.18–28.10.18	POW
Lt.	Katzenstein	Kurt	11.08.17–10.01.18	Aflup F
			26.04.18–11.11.18	
Lt.	Kaus	Erich	02.12.17–31.05.18	wounded
Uffz.	Liebert	Emil	09.09.17–03.01.18	killed in crash
Lt.	Maier	Reinhold	30.12.17–28.01.18	Hospital
			06.05.18–19.10.18	died after crash
Uffz.	Marczinke	Max	19.03.18–30.03.18	POW
Uffz.	Marszinski	Paul	05.05.18–22.7.18	KIA

Rank	Last Name	First Name	Period of Service	Remarks
Lt.	von der Marwitz	Hans-Georg	18.04.17–17.4.18	Staffelführer
Lt.	Morhau		23.01.18–10.02.18	Hospital
Lt.	Nernst	Gustav	20.01.17–21.04.17	KIA
Lt.	Oberländer	Hans	15.05.17–11.06.18	FEA 14
			17.09.18–11.11.18	
Oblt.	Preissler	Kurt	22.10.17–19.3.18	Staffelführer
Lt.	Pruefer	Ewald	23.10.18–11.11.18	
Uffz.	Rody	Hans	19.01.17–14.03.17	killed in crash
Lt.	von Schell	Hans	16.01.17–03.10.17	Aflup 6
Vzfw.	Schiebler	Arthur	19.04.18–27.05.18	KIA
Oblt.	Schlieter	Hans	05.06.17–20.08.17	Aflup D
			01.11.17–06.02.18	Jasta 70
Gefr.	Schneider	Heinrich	19.01.17–15.02.17	POW
Lt.	Seewald		14.12.17–20.01.18	Jasta 54
Lt.	Seitz	Oskar	16.04.17–20.07.17	Aflup 6
Flg.	Steiner	Philipp	17.09.18–11.11.18	
Vzfw.	Wagner	Wilhelm	07.04.18–27.09.18	Aflup 6
Lt.	Weltz	Karl	24.06.17–15.12.17	B.L.A.
Vzfw.	Willmann	Max	18.07.17–21.09.17	Aflup 6

4.Offz.z.b.V.

Rank	Last Name	First Name	Period of Service	Remarks
Lt.	Schnorr	Douglas	20.01.17–15.02.18	Jasta 38
Oblt.	Preissler	Kurt	15.02.18–16.4.18	JG II
Lt.	Siempelkamp	Egon	11.05.18–29.08.18	Hospital
Lt.	Maier	Reinhold	01.09.18–19.10.18	died after crash

Aircraft Documented as Having Flown With Jagdstaffel 30			
Military Serial	Receipt*	Dispatch/Final Mention	Remarks
Halberstadt D V			
D.412/16	08.02.1917	?	
D.414/16	13.02.1917	?	
D.420/16	08.02.1917	15.02.1917	Gefr. Schneider POW
Albatros D III			
D.2038/16	25.02.1917	17.03.1917	Aircraft of Lt. Nernst; Uffz. Rody killed
D.2051/16	25.02.1917	07.07.1917	Aircraft of Oblt. Bethge till end of March; destroyed behind German lines
D.2054/16	25.02.1917	July/August 1917	Delivered to Aflup 6; Aircraft of Lt. Brügman
D.2084/16	14.03.1917	?	
D.2098/16	14.03.1917	?	
D.2122/16	14.03.1917	July/August 1917	Delivered to Aflup 6
D.2124/16	14.03.1917	?	Aircraft of Lt. Nernst (March/April 1917)
D.2126/16	27.03.1917	?	Crash landing, aircraft of Vfw. Heiligers
D.2140/16	14.03.1917	?	Airctaft of Lt. Erbguth
D.2141/16	14.03.1917	July/August 1917	Delivered to Aflup 6
D.2147/16	27.03.1917	21.04.1917	Aircraft of Oblt. Bethge; Lt. Nernst killed
D.2304/16	12.04.1917	July/August 1917	Delivered to Aflup 6; Aircraft of Lt. Seitz
D.2305/16	11.04.1917	?	Aircraft of Lt. von Schell
D.2306/16	12.04.1917	(31.08.1917)	
D.605/17	13.05.1917	(31.08.1917)	
D.760/17	21.05.1917	July/August 1917	Badly damaged while flipping over
D.767/17	26.04.1917	?	
D.791/17	01.06.1917	(02.12.1917)	
D.799/17	21.05.1917	14.10.1917	
Albatros D V			
D.1012/17	22.05.1917	July 1917	Back to Aflup 6 (Kofl 6, Nr. 3597 VI); aircraft of Lt. Erbguth
D.1016/17	21.05.1917	July 1917	Back to Aflup 6 (Kofl 6, Nr. 3597 VI); aircraft of Lt. von der Marwitz;
D.1017/17	21.05.1917	?	
D.1023/17	22.05.1917	July 1917	Back to Aflup 6 (Kofl 6, Nr. 3597 VI)
D.2047/17	03.07.1917	?	
D.2051/17	03.07.1917	(07.12.1917)	
D.2126/17	12.07.1917	(19.02.1918)	Aircraft of Lt. Holthusen ("Corkscrew")
D.2140/17	12.07.1917	(31.08.1917)	Aircraft of Lt. Fuchs
D.2150/17	01.08.1917	15.08.1917	Lt. Brügman killed
D.2177/17	21.07.1917	(31.08.1917)	
D.2186/17	22.07.1917	(31.08.1917)	Aircraft of Vzfw. Heiligers

Aircraft Documented as Having Flown With Jagdstaffel 30			
Military Serial	**Receipt***	**Dispatch/Final Mention**	**Remarks**
D.2191/17	20.07.1917	12.08.1917	Lt. von Bertrab POW
D.2196/17	?	20.11.1917	Vzfw. Heiligers killed
D.2197/17	21.07.1917	(31.08.1917)	
D.2198/17	01.08.1917	(31.08.1917)	
D.2199/17	01.08.1917	(15.10.1917)	
D.2203/17	22.07.1917	(05.01.1918)	
D.2205/17	21.07.1917	(31.08.1917)	
D.2207/17	21.07.1917	(31.08.1917)	
D.4420/17	19.08.1917	(31.08.1917)	Aircraft of Lt. Weltz; 1918 at FEA 13
D.4422/17	19.08.1917	(16.02.1917)	
D.4477/17	05.09.1917	?	
D.4501/17	11.09.1917	?	
D.4514/17	20.09.1917	(24.12.1917)	
Pfalz D III			
D.4078/17	21.10.1917	?	
D.4079/17	22.10.1917	?	
D.4080/17	27.10.1917	?	
D.4085/17	22.10.1917	?	
D.4086/17	27.10.1917	?	
D.4087/17	21.10.1917	(24.10.1917)	
D.4093/17	17.12.1917	?	
D.4097/17	17.12.1917	?	
Pfalz D IIIa			
D.4202/17	?	?	
D.4203/17	06.12.1917	?	Aircraft of Oblt. Bethge
D.5888/17	(16.02.1918)	17.03.1918	Oblt. Bethge killed
D.5891/17	(06.01.1918)	(01.06.1918)	Aircraft of Lt. Holthusen
D.5947/17	?	?	
D.5981/17	(12.03.1918)	?	
D.5983/17	(Feb.1918)	?	
D.5985/17	(11.03.1918)	(12.03.1918)	
D.5986/17	(12.03.1918)	(11.05.1918)	
D.8078/17	?	30.03.1918	Uffz. Marczinke POW[1]
D.8233/17	?	?	
Fokker D VII			
D.370/18	(01.07.1918)	(23.09.1918)	Aircraft of Lt.Holthusen
D.541/18	(28.07.1918)		Test aircraft (wooden fuselage)
Pfalz D.XII			
D.1488/18	(04.08.1918)	?	Wine red fuselage (Lt. von der Marwitz?)
DFW C.V			
C.6104/17	(Feb.1918)		Staffel hack aircraft

[1]US/UK sources indicate D.8278/17. *Dates in parentheses indicate first or last known mention with Staffel.

End Notes

Prolog

1. Otto Fuchs, *Wir Flieger*, p. 104.

1. December 1916 – May 1917

1. Otto Fuchs, Wir Flieger, p. 129.
2. Interview Bruno Schmäling with Otto Fuchs, 1976.
3. The weekly report of the Kommandeur identifies the airplane in which Lt. von Schell crashed as Albatros D V D.414/17. Here a typographical error is involved, as there was no Albatros D V with the military serial number D.414/17! It is very probable that the machine concerned was Albatros D V D.1014/17.
4. Kette = group of three aircrafts. The authors decided to use the German word "Kette" instead of "flight" which described not exactly a formation of three fighter aircrafts.
5. Walter Zuerl, *Pour le mérite-Flieger*, pp. 496-497.
6. Military Science Division of the Luftwaffe, Englische Flugtaktik, p. 97. This draft prepared for official use by the German Luftwaffe (1933) deals with the combat experiences against the English air service during the battle in Flanders in July 1917.
7. *Englische Flugtaktik*, p. 98.
8. This was the first aerial victory at night of a fighter pilot. The first aerial victory at night was achieved by a two-seater crew in February 1917.
9. Personnel file of Oskar Seitz, Bavarian Main State Archive, Dept. 4, OP 12233.
10. War Diary of the Royal Flying Corps, compiled by Les Rogers.
11. Report Paul Erbguth, owned by Erbguth family
12. A.O.K. 6 weekly report for the week of 25 to 31 May 1917. Bayerisches Hauptstaatsarchiv, Abteilung IV, München (Bavarian Main State Archive, Department 4, Munich.
13. Report of the Flakgruppe-Kommando in the Flugmeldebuch (aircraft spotting book) of the German 6th Army on April 3, 1917: "Nieuport went down steeply after aerial combat with Albatros with green upper wing and green spinner over Esquerchin and landed in the vicinity of the airfield at Douai, Bayerisches Hauptstaatsarchiv Abteilung 4, München (Bavarian Main State Archive, Department IV, Munich).
14. Various former pilots from Jagdstaffeln confirmed independently of one another that letter and number markings were not usually done in black unless the black letter was bordered white. Red seems to be a favourite color for letters and numbers, but also light blue and yellow were used. This was due to the good recognizability from a great distance, both for other pilots as well as for observers on the ground. The more striking the paint scheme of a fighter plane was, the greater the chance also that a victory would be clearly confirmed by observers on the ground. Red letters or numbers also stood out better from the black national insignia.
15. Personal information from Hans Holthusen.

2. June 1917 – August 1917

1. Personnel file of Otto Fuchs (OP 24636), Bavarian Main State Archive, Dept. 4, Munich.
2. Otto Fuchs personnel file.
3. Otto Fuchs, *Wir Flieger*, p. 131. With subsequent corrections and additions by him.
   It should be noted that his statements reveal here a slight error in memory: at the beginning of June 1917 Karl was already a "Kanone" with over 20 victories, but he did not receive the Pour le Mérite until June 14th, by which time Fuchs had already been with Jagdstaffel 30 for a week. It shoud be noted, that the term "ace" was completely unknown in the German air service.
4. Otto Fuchs, *Wir Flieger*, p.128, with subsequent corrections and additions by him.
5. Ibid., p. 133 ff., with subsequent corrections and additions by him.
6. Ibid., p. 140, with subsequent corrections and additions by him.
7. Personal statement of Otto Fuchs to Bruno Schmäling, 1976.
8. Personal statement of Otto Fuchs to Bruno Schmäling, 1976.
9. Otto Fuchs, Wir Flieger, p. 136 ff.
10. Ibid., p. 135.
11. Ibid., p. 137 ff., with subsequent corrections and additions by him.
12. Otto Fuchs, *Wir Flieger*, p. 162 ff.
13. Personal statement of Otto Fuchs to Bruno Schmäling, 1976.
14. Combat Report of 2/Lt. Albert Earl Godfrey, 40 Sqn. RFC.
15. Combat Report of 2/Lt Gerard Bruce, 40 Sqn. RFC.
16. AOK 6 weekly aviation report for the week of June 1 to 28, 1918, Bavarian Main State Archive, Munich.
17. Oskar Seitz personnel file OP 12233, Bavarian Main State Archive, Munich.
18. Ibid.
19. Otto Fuchs, *Wir Flieger*, p. 146.
20. Ibid., p. 150 ff.
21. Ibid., p. 151, with subsequent additions by him.

22. Ibid., p. 152, with subsequent additions by him.
23. Personal statement of Otto Fuchs to Bruno Schmäling, 1977.
24. Logbook of Kurt Katzenstein, Deutsches Technikmuseum, Berlin.
25. Ibid.
26. Frederick Oughton, The Personal Diary of "Mick" Mannock, p. 127.
27. Personal statement of Otto Fuchs, 1976.
28. Logbook of Kurt Katzenstein, Deutsches Technikmuseum, Berlin.
29. Ibid. Unfortunately, the logbook of Kurt Katzenstein available at the Deutsches Technikmuseum only covers the period of August 10, 1917 to August 20, 1917.
30. Personal statement of Otto Fuchs to Bruno Schmäling, 1977.
31. The representation of the painting of the aircraft according to the statements of Otto Fuchs, who as a painter had a good recollection of colors.
32. Otto Fuchs, *Wir Flieger,* p. 140.
33. Personal statement of Otto Fuchs, 1977.
34 Otto Fuchs, *Wir Flieger,* p. 141.
35. Personal statement of Otto Fuchs to Bruno Schmäling, 1977.

3. September 1917 – October 1917

1. Aircraft inventory list of Armee-Flug-Park 6. Compiled by Reinhard Zankl. The inventory for the Armeeflugpark identifies 12 Albatros D Vs as being with the Staffel on Augst 31, 1917. In addition, according to the logbook of Lt. Katzenstein at least Albatros D III D.791/17 and D.799/17 were at the Staffel. According to the logbook of Hans Holthusen both airplanes were still at the Staffel in October 1917.
2. Personal statement of Otto Fuchs to Bruno Schmäling, 1977.
3. Report of Lt. Paul Erbguth in the private collection of the Erbguth family.
4. A.O.K. 6, Bund 304. Bavarian Main State Archive, Munich.
5. Logbook of Lt. Hans Holthusen.
6. Ibid.
7. Otto Fuchs, Wir Flieger, p. 169 ff., with subsequent corrections and additions by him. In the report of Flak-Gruppe 2 it is stated regarding this aerial combat:
   DFW attacked by three Sopwiths. 10:45 DFW shot down by a Sopwith, ½ minute later Sopwith shot down by an Albatros. Both machines lie close together near Harnes. DFW from Flieger-Abteilung (A)253, the occupants were brought to the main dressing station near Dourges. (Flugmeldebuch der deutschen 6. Armee, Bavarian Main State Archive, Munich).
8. War Diary of the Royal Flying Corps, compiled by Les Rogers from correspondence with Dr. Gustav Bock.
9. J. Beedle, *43 Squadron: The History of the Fighting Cocks, 1916–1966*, p. 64.
10. Otto Fuchs, *Wir Flieger*, p. 171 ff., with subsequent corrections and additions by him.
11. Ibid., p. 178 ff., with subsequent corrections and additions by him.
12. Ibid., p. 188 ff., with subsequent corrections and additions by him.
13. Ibid., p. 158.
14. Ibid., p. 160 ff.
15. Ibid., p. 188 ff. with subsequent corrections and additions by him.
16. Otto Fuchs, *Wir Flieger*, unpublished portion of manuscript, Schmäling collection.
17. Otto Fuchs, *Wir Flieger*, p.195 ff., with subsequent corrections and additions by him.
18. Logbook of Hans Holthusen, 1917–1918, copy in Schmäling collection.
19. Personal statement of Otto Fuchs. Unfortunately, there is no photo of this machine available, although Otto Fuchs was certain that he did have one. After the war he painted the same fox on his own self-constructed airplane, which incidentally also received an emerald-green paint job. However, the fox here was white.
20. Personal statement of Otto Fuchs.
21. Letter from Hans Holthusen and personal statement of Hans Holthusen. In his letter of August 30, 1980 to one of the authors it is stated: "My airplanes were ringed in red and white and probably for this reason had the nickname 'corkscrew'... Up to my Fokker." Upon further inquiry, it was determined that it concerned the Albatros D V and Pfalz D III. A drawing was prepared with Herr Holthusen's help and forms the basis for the illustration in this publication.
22. Personal statement of Otto Fuchs to Bruno Schmäling, 1977.

4. November 1917 – February 1918

1. Otto Fuchs, *Wir Flieger*, unpublished portion of manuscript, Schmäling collection. The last conversation with Hans Bethge was among the parts of his book which were stricken by the editor in 1933. Hans Bethge's doubt in a military victory and his open criticism of the military leadership did not fit in with the politically desirable depiction at this time.
2. Ibid.
3. Letter from Erich Kaus to Dr. Gustav Bock, 1973, Bock collection.

4. Georg Richter, *Der königlich sächsische Militär St. Heinrichs Orden 1736–1918.*
5. Letter from Erich Kaus to Dr. Gustav Bock, 1973, Bock collection.
6. Logbook of Lt. Hans Holthusen, 1917–1918.
7. Combat report of 2/Lt. Percy Jack Clayson, 1 Sqn. RFC, Bock collection.
8. Logbook of Hans Holthusen. The DFW C.V C.6401/17 was delivered to Flg. Abt. (A) 235 on November 18, 1917. On December 31st it was handed over to Aflup 6 and from there found its way to Jagdstaffel 30.
9. Information from Alex Imrie, confirmed by Erich Kaus and Hans Holthusen.
10. G-Report 134, Imrie collection. In the G-Report can be read: Fuselage painted yellow with black line running diagonally across it. Rudder and empennage also yellow having black markings on it. Top and LH (left hand) bottom planes mauve and brown above and light blue below. RH (right hand) bottom plane lozenge fabric.
11. The colors of the flag of the Kingdom of Württemberg were black and red, the colors in the coat of arms yellow and black. In this connection Royal Württemberg Jagdstaffel 28 used the colors yellow and black as a Staffel marking, while Royal Württemberg Jagdstaffel 64 used the colors black and red as a Staffel marking. In addition, yellow and black were the colors of the district of Esslingen, to which Reinhold Maier's hometown belonged.
12. Personal statement of Erich Kaus. Until 1866 Hannover was an independent Kingdom. After the war with Prussia it become a province of Prussia, but like many other people from Hannover Erich Kaus still felt like a member of the Kingdom of Hannover and not a Prussian.
13. Statement of Alex Imrie based on his conversation notes with Erich Kaus.
14. Otto Busche was transferred to Jagdstaffel 30 on September 24, 1917. On February 15, 1918, he was transferred to Armee-Flug-Park 6 and then further to Flg. Abt. 5.

5. March 1918 – June 1918

1. AOK 6 weekly aviation report for the week of March 8 to 14, 1918, Bavarian Main State Archive, Munich.
2. Walter Zuerl, *Pour le mérite-Flieger*, pp. 494–498.
3. AOK 6 weekly aviation report for the week of April 12 to 18, 1918, Bavarian Main State Archive, Munich
4. AOK 6 weekly aviation report for the week of May 10 to 16, 1918, Bavarian Main State Archive, Munich.
5. Combat Report of 40 Squadron RAF (Dallas, Lewis, Hind).
6. Letter from Erich Kaus to Dr. Gustav Bock, 1973.
7. Personal statement of Erich Kaus to Alex Imrie.
8. AOK 6 weekly aviation report for the week of May 31 to June 6, 1918, Bavarian Main State Archive, Munich.
9. Personal statement of Hans Holthusen to Bruno Schmäling.
10. Personal statement of Erich Kaus to Bruno Schmäling

6. July 1918 – November 1918

1. Logbook of Hans Holthusen, 1917–1918.
2. Personal statement of Hans Holthusen to Bruno Schmäling.
3. Combat report of Lt. Gilbert John Strange, 40 Squadron RAF, Bock archive.
4. AOK 6 weekly aviation report for the week of July 19 to 25, 1918, Bavarian Main State Archive, Munich.
5. Logbook of Hans Holthusen, 1917–1918.
6. Ibid.
7. Ibid.
8. AOK 6 weekly aviation report for the week of August 9 to 15, 1918, Bavarian Main State Archive, Munich.
9. Bill Puglisi, ed., "Raesch of Jasta 43," *The Cross & Cockade Journal*, Vol 8, No. 4.
10. AOK 6 weekly aviation report for the week of August 16 to 22, 1918, Bavarian Main State Archive, Munich.
11. Bill Puglisi, ed., "Raesch of Jasta 43," *The Cross & Cockade Journal*, Vol 8, No. 4.
12. Kommandeur der Flieger 6. Armee, Befehl Nr. 69726/1016, Akte ILUFT 4, Bavarian Main State Archive, Munich.
13. AOK 6 Kommandeur der Flieger Abt. I/N.O./ III Nr. 1834 geh. Bavarian Main State Archive, Munich.
14. Report Flg. Philip Steiner, private property
15. Letter from Hans Holthusen of August 30, 1980: "My Fokker had an orange-colored fuselage and a black nose. Marwitz had all his aircraft painted wine red. One Fokker had his coat of arms on the fuselage, that was just before I was transferred. I can remember the witch on the broom you mentioned. There were yet other insignia, but I can no longer recall them."
16. Hans Holthusen personal statement to Bruno Schmäling, 1980.
17. Letter from Hans Holthusen of August 30, 1980.
18. Correspondence of Erich Tornuss with August Hartmann, 1968, Bock archive.

Excerpt from the War Diary of Royal Prussian Jagdstaffel 30

14.12.16 Beginning of the formation of the Staffel at FEA 11 in Breslau.

15.01.17 Oblt. Hans Bethge, previously of Jagdstaffel 1, is named Staffelführer.

16.01.17 Lt. Hans von Schell transferred from Jagdstaffelschule I to the Staffel and temporarily appointed as Special Duty Officer.

18.01.17 Vzfw. Gustav Eimbeck, previously Jagdstaffel 13, is transferred to the Staffel.
Uffz. Bruno Beyer, previously Jagdstaffel 9, is transferred to the Staffel.

19.01.17 Uffz. Hans Rody and Gefr. Heinrich Schneider, previously Jagdstaffel 21, are transferred to the Staffel.

20.01.17 Lt. Douglas Schnorr arrives at the Staffel and assumes the duties of the Special Duty Officer.
Lt. Gustav Nernst, previously Jagdstaffel 10, is transferred to the Staffel.
Uffz. Josef Heiligers is transferred from Armeeflugpark 6 to the Staffel.

21.01.17 First day of mobilization for the Staffel.

22.01.17 Beginning of railway transport of the Staffel from Breslau to the Western Front.

25.01.17 Staffel moves to Phalempin airfield in the 6. Armee sector and immediately begins construction and expansion of the airfield facilities.

15.02.17 Gefr. Heinrich Schneider takes off for a practice flight in Halberstadt D III D.420/16 and does not return from this flight. According to observations from the front, he made an emergency landing over Ploegsteert Wood in enemy territory after being fired upon by anti-aircraft guns and fell into English captivity.

25.02.17 Lt. Hans von Schell flipped over while landing at Phalempin airfield and is sent to a hospital with severe injuries.

01.03.17 First operational flight of the Staffel.

06.03.17 Lt. Joachim von Bertrab is transferred from Armeeflugpark A to the Staffel.

10.03.17 Lt. Paul Erbguth, previously of Schutzstaffel 21, arrives at the Staffel.

17.03.17 Uffz. Hans Rody crashed fatally after a collision in the air during a practice flight in the vicinity of Phalempin airfield.

28.03.17 Oblt. Hans Bethge flying Albatros D III D.2051/16 shoots down a Nieuport at 10:40, which falls apart over our territory to the west of Bouvines. Afterwards he must make an emergency landing with hits in his aircraft, during which the machine is damaged. Downed aircraft is confirmed by the Kogenluft as his 4th aerial victory.
According to available records the downed machine was Nieuport A6615 of 1 Squadron RFC. Pilot 2/Lt. H. Welch was killed.

30.03.17 Lt. Heinrich Brügman, previously Flg. Abt. (A)240, arrives at the Staffel.

03.04.17 Lt. Gustav Nernst flying Albatros D III D.2124/16 forces a Nieuport to land over our territory south of Douai near Esquerchin at 14:42. Downed aircraft is confirmed by the Kogenluft as his 3rd aerial victory.

According to available records the downed machine was Nieuport A6674 of 40 Squadron RFC. Pilot 2/Lt. S. A. Sharpe fell into captivity unwounded.

05.04.17 Lt. Gustav Nernst Albatros D III D.2124/16 forces a Sopwith two-seater to land in our territory west of Rouvroy at 12:05. Downed aircraft is confirmed by the Kogenluft as his second aerial victory.
According to available records the downed machine was Sopwith 1½-Strutter A1073 from 43 Squadron RFC. Pilot 2/Lt. C. P. Thornton fell into captivity unwounded, observer 2/Lt. H. D. Blackburn died.

06.04.17 Lt. Joachim von Bertrab shoots down a Sopwith out of a formation at 07:40 over our territory between Ath and Bouvignies, after a lengthy pursuit he shoots down a second Sopwith from the same formation over our territory southeast of Leuze near Ramegnies at 08:15. Downed aircraft are confirmed by the Kogenluft as his 1st and 2nd aerial victories.
According to available records the two downed machines were Martinsyde G.100s 7465 and 7478 from 27 Squadron RFC. Pilots 2/Lt. J. R. S. Proud and Lt. J. H. B. Wedderspoon died.

Lt. Joachim von Bertrab shoots down two Sopwiths from another formation at 10:42 over our own territory between Pecq and Obigies. The downed machines are confirmed by the Kogenluft as his 3rd and 4th aerial victories.
According to available records the two machines brought down were Sopwith 1½-Strutters A2381 and A1093 from 45 Squadron RFC. The crews 2/Lt. C. St.G. Campbell (pilot) and Capt. D. W. Edwards, MC (observer) as well as 2/Lt. J. A. Marshall (pilot) and 2/Lt. F. G. Truscott, MC died.

Oblt. Hans Bethge shoots down a Sopwith from the same formation in flames at 10:48 over our territory at the Nechin railway station near Tournai. The downed aircraft is confirmed by the Kogenluft as his 5th aerial victory.
According to available records the downed machine was Sopwith 1½-Strutter 7806 of 45 Squadron RFC. The crew, 2/Lt. J. E. Blake (pilot) and Capt. W. S. Brayshay (observer), died.

12.04.17 Vzfw. Gustav Eimbeck is transferred to Armeeflugpark 6.

16.04.17 Lt. Oskar Seitz, previously of Fl. Abt. (A)292, is transferred to the Staffel.

18.04.17 Lt. Hans-Georg von der Marwitz is transferred from Jagdstaffelschule I to the Staffel.

20.04.17 Oblt. Hans Bethge goes on leave for four weeks; Lt. Paul Erbguth assumes acting Staffelführer of the Staffel.

21.04.17 Lt. Gustav Nernst in Albatros D III D.2147/16 is rammed by Lt. Oskar Seitz at 18:45 in combat with F.E.s and Triplanes and thereafter crashes fatally in enemy territory near Arras. Lt. Seitz is able to glide to a smooth emergency landing with a damaged machine in enemy territory near Gavrelle and make his way to our lines on foot.
According to available records the opponents in this aerial combat were F.E. 2bs from 25 Squadron RFC and Sopwith Triplanes from 8 Squadron RNAS. The crew of 2/Lt. R. G. Malcolm and 2/Lt. J. B. Weir in F.E. 2b 4839 and F/Cdr. A. R. Arnold in Sopwith Triplane N5458 reported the crash of Lt. Nernst as a combined victory. F/Cdr. Arnold furthermore reported the downing of a second Albatros D III in flames and F/S/Lt. R. A. Little in Sopwith Triplane N5469 the downing of a third Albatros D III, the impact of which on the ground was observed.

Lt. Oskar Seitz receives 14 days' recuperative leave.

22.04.17 Uffz. Bruno Beyer is awarded the pilot's badge.

24.04.17 Lt. Joachim von Bertrab and Uffz. Josef Heiligers each shoot down a B.E. at 07:45 over our territory between Douai and Valenciennes. Both victories are not confirmed.

26.04.17 Uffz. Bruno Beyer receives the Iron Cross, 2nd Class.

Lt. Paul Erbguth forces a B.E. to land over our territory southeast of Haisnes at 18:28. Downed

aircraft is confirmed by the Kogenluft as his 1st aerial victory.
According to available records the downed machine was B.E. 2e 5870 of 10 Squadron RFC. The crew of 2/Lt. F. Roux (pilot) and 2/Lt. H. J. Prince (observer) fell into captivity wounded.

01.05.17 Vzfw. Franz Bucher arrives at the Staffel from Jagdstaffelschule I.

Lt. Paul Erbguth and Lt. Heinrich Brügman each shoot down a Sopwith over our territory between Douai and Hénin-Liètard. Both victories were not confirmed.

05.05.17 Lt. Oskar Seitz returns to the Staffel from home leave.

07.05.17 Vzfw. Franz Bucher shoots down a Sopwith over enemy territory near La Bassée. The victory was not confirmed.

13.05.17 Lt. Hans-Georg von der Marwitz shoots down a captive balloon in flames at 14:40 over enemy territory between Ypres and Dixmude. The downed balloon is confirmed by the Kogenluft as his 2nd aerial victory.

15.05.17 Vzfw. Hans Oberländer is transferred from Jagdstaffelschule I to the Staffel via Armeeflugpark 6.
Lt. Joachim von Bertrab forces an F.E. to land at 8:30 over our own territory between Quesnoy and Wambrechies. Downed aircraft confirmed as his fifth aerial victory by the Kogenluft.
According to available records the downed machine was F.E. 2d A6446 from 20 Squadron RFC. The crew of S/Lt. E. J. Grout (pilot) and 2/AM A. Tyrell (machine-gunner) fell into captivity unwounded.

20.05.17 Oblt. Hans Bethge returns from leave and again assumes leadership of the Staffel.

Gefr. Josef Funk is transferred from Armeeflugpark 6 to the Staffel.

21.05.17 Lt. Hans von Schell crashes while landing at Phalempin airfield and is sent to the hospital badly injured.

26.05.17 Oblt. Hans Bethge forces a Nieuport to land at 20:55 over our own territory east of Esquerchin. Downed aircraft confirmed by the Kogenluft as his 6th aerial victory.
According to available records the downed machine was Nieuport B1626 from 29 Squadron RFC. Pilot 2/Lt. G. M. Robertson falls into captivity wounded.

31.05.17 Gefr. Josef Funk shoots down a captive balloon in flames at 10:10 over enemy territory near Nieppe. Downed balloon confirmed by the Kogenluft as his 1st aerial victory.

01.06.17 From 25 till 31 May the Staffel undertook 113 operational flights for a combined duration of 105 flight hours in the course of which it engaged in 34 aerial combats.

Oblt. Hans Bethge shoots down an F.E. 8 at 21:30 over our own territory northeast of Warneton. Downed aircraft confirmed by the Kogenluft as his 7th aerial victory.
According to available records the downed machine was F.E. 8 A4887 from 41 Squadron RFC. Pilot 2/Lt. P. C. S. O'Longan died.

05.06.17 Lt. Hans Schlieter and Lt. Otto Fuchs are transferred from Jagdstaffelschule I via Armeeflugpark 6 to the Staffel.

Oblt. Hans Bethge forces a Spad single-seater to land at 12:45 over our territory east of Frélinghien. Downed aircraft confirmed by the Kogenluft as his 8th aerial victory.
According to available records the downed machine was Spad VII A6747 from 19 Squadron RFC. Pilot 2/Lt. C. D. Grierson fell into captivity unwounded.

07.06.17 Uffz. Josef Heiligers shoots down an F.E. at 8:00 over our territory west of Menin near Koelberg. Downed aircraft confirmed by the Kogenluft as his 1st aerial victory.
According to available records the downed machine was F.E.2d A1957 of 25 Squadron RFC. Pilot 2/Lt. G. H. Pollard fell into captivity badly wounded, observer 2/Lt. F. S. Ferriman died.

Vzfw. Hans Oberländer forces a Sopwith single-seater to land at 11:45 on our territory south of Roulers. Downed aircraft confirmed by the Kogenluft as his 2nd aerial victory.
According to available records the downed machine was Sopwith Pup A7314 from 66 Squadron RFC. Pilot 2/Lt. R. M. Marsh fell into captivity unwounded.

08.06.17 From 1 to 7 June the Staffel undertook a total of 117 operational flights with a combined duration of 130 flight hours, in the course of which it engaged in 55 aerial combats.

Vzfw. Franz Bucher is shot down in aerial combat and crashes fatally in the vicinity of Wervicq.
According to available records the opponents were possibly F.E. 2ds of 20 Squadron RFC, which in a combat with six Albatros D IIIs reported an aircraft brought down between Comines and Becelaere by the crew of 2/Lt. W. Durrand (pilot) and Sgt. E. Sayers (machine-gunner) in FE2d A1965.

15.06.17 From 8 to 14 June undertook a total of 98 operational flights with a combined duration of 100 flight hours, in the course of which it engaged in 32 aerial combats.

16.06.17 Lt. Hans Forstmann is transferred from Jagdstaffelschule I to the Staffel.

21.06.17 Lt. Otto Fuchs shoots down a Bristol Fighter in flames at 8:03 over our territory northwest of Vitry-en-Artois. Downed aircraft is confirmed by the Kogenluft as his 1st aerial victory.
According to available records the downed machine was Bristol F2b A7139 from 11 Squadron RFC. Pilot 2/Lt. D. C. H. MacBrayne died, machine-gunner W. Mollison fell into captivity unwounded.

22.06.17 From 15 to 21 June the Staffel undertook a total of 129 operational flights with a combined flight duration of 132 flight hours, in the course of which it engaged in 18 aerial combats.

24.06.17 Lt. Karl Weltz, previously of Kampfgeschwader 1/5, is transferred to the Staffel.

Uffz. Josef Heiligers shoots down a Nieuport at 19:40 over our territory southwest of the Beaumont railway station. Downed aircraft is confirmed by the Kogenluft as his 2nd aerial victory.
According to available records the downed machine was Nieuport B1607 from 29 Squadron RFC. Pilot Capt. W. P. Holt died.

29.06.17 From 22 to 28 July the Staffel undertook a total of 132 operational flights with a combined duration of 142 flight hours, in the course of which it engaged in 36 aerial combats.

02.07.17 Lt. Hans Forstmann is shot down at 11:30 in aerial combat with 3 Nieuports and crashes fatally by the canal near Dourges.
According to available records the opponents were Nieuports of 40 Squadron RFC, who reported the downing of an Albatros DIII in aerial combat at 10:30 north of Douai by 2/Lt. A. E. Godfrey in Nieuport B1684. 20 minutes later Lt. G. B. Crole in Nieuport B1552 reported downing an Albatros D III in flames northwest of Douai.

06.07.17 From 29 June to 5 July the Staffel undertook a total of 73 operational flights with a combined flight duration of 73 flight hours, in the course of which it engaged in 16 aerial combats.

09.07.17 Uffz. Bruno Beyer is transferred via Armeeflugpark 6 to Fl. Abt. (A)235.

10.07.17 Gefr. Josef Funk is awarded the Golden Military Order of Merit.

13.07.17 From 6 to 12 July the Staffel undertook a total of 67 operational flights with a combined flight duration of 70 flight hours, in the course of which it engaged in 18 aerial combats.

Uffz. Josef Heiligers forces a Sopwith to land in enemy territory near Annequin at 21:25. Downed aircraft is confirmed by the Kogenluft as "forced to land on the other side of the lines."

14.07.17 Uffz. Josef Heiligers forces an Avro single-seater to land in our territory between Evin-Malmaison and Le Forest at 08:00. Downed aircraft confirmed by the Kogenluft as his 3rd aerial victory.
According to available records the downed machine was Martinsyde G.100 A1572 from 27 Squadron RFC. Pilot 2/Lt. G. H. Palmer fell into captivity unwounded.

15.07.17 Oblt. Hans Bethge shoots down a Martinsyde in flames at 21:20 over our own territory between Loison and Hannes. Downed aircraft confirmed by the Kogenluft as his 9th aerial victory.
According to available records the downed machine was Sopwith Camel B3758 from 8 Squadron RNAS. Pilot F/S/Lt. F. Bray died.

18.07.17 Vzfw. Max Willmann, previously of Fl. Abt. (A)292, is transferred via Armeeflugpark 6 to the Staffel.

20.07.17 Lt. Oskar Seitz is transferred via Armeeflugpark 6 to Fl. Abt. (A)292.

From 13 to 19 July the Staffel undertook a total of 88 operational flights with a combined flight duration of 96 flight hours, in the course of which it engaged in 17 aerial combats.

Oblt. Hans Bethge shoots down a Sopwith single-seater at 21:25 over enemy territory near Méricourt. Downed aircraft confirmed by the Kogenluft as his 10th aerial victory.

21.07.17 Gefr. Josef Funk receives the Iron Cross 1st Class.

23.07.17 Uffz. Josef Heiligers receives the Iron Cross 1st Class.
Lt. Hans Schlieter is sent to a hospital with an illnes.

27.07.17 From 20 to 26 July the Staffel undertook a total of 106 operational flights with combined flight duration of 100 flight hours, in the course of which it engaged in 22 aerial combats.

03.08.17 From 27 July to 2 August the Staffel undertook a total of 78 operational flights with a combined flight duration of 80 flight hours, in the course of which it engaged in 27 aerial combats.

10.08.17 From 3 to 9 August the Staffel undertook a total of 19 operational flights with a combined flight duration of 16 flight hours, in the course of which it engaged in no aerial combats.

11.08.17 Lt. Kurt Katzenstein is transferred from Jagdstaffelschule I to the Staffel.

12.08.17 Lt. Joachim von Bertrab took off in Albatros D V D.2191/17 for an attack on the captive balloon near Souchez and did not return. According to observations from the front was attacked by enemy fighters and forced to land at 16:00 at Farbus Wood. Fell into English captivity.
According to available records the opponent was 2/Lt. E. Mannock from 40 Squadron RFC in Nieuport B3554.

Uffz. Josef Heiligers forces a Nieuport to land at 20:45 in our own territory east of Courrières. Victory was at first contested by Lt. Ernst Udet from Jagdstaffel 37, which likewise took part in the aerial combat, and through arbitration was awarded to Uffz. Heiligers. Downed aircraft was confirmed by the Kogenluft as his 4th aerial victory.
According to available records the downed machine was Nieuport A6771 from 40 Squadron RFC. Pilot 2/Lt. C. W. Cullen fell into captivity unwounded.

14.08.17 Lt. Heinrich Brügman shoots down a Martinsyde which falls apart at 20:25 over our territory near Pont-à-Vendin. Victory was contested by Ernst Udet of Jagdstaffel 37, which participated in the combat, and it was granted to him.

15.08.17 Lt. Heinrich Brügman is shot down in Albatros D V D.2051/17 at 11:15 in aerial combat and landed with a severe abdominal wound near Douvrin. After being recovered and receiving emergency medical attention, he dies around 14:00 on the transport to a rear area hospital.
According to available records the opponents were probably DH4s from 25 Squadron RFC, who in a combat with 12 Albatros D Vs report the downing of an enemy plane in flames by the crew Capt. J. Morris (pilot) and Lt. D. Burgess (observer).

Oblt. Hans Bethge shoots down a Nieuport at 20:45 which falls apart over our territory north of Loison near Annay. Downed aircraft confirmed by the Kogenluft as his 11th aerial victory.
According to available records the downed machine was Nieuport B1682 from 40 Squadron RFC. Pilot Capt. W. G. Pender died.

17.08.17 From 10 to 16 August the Staffel undertook a total of 207 operational flights with a combined flight duration of 198 flight hours, in the course of which it engaged in 46 aerial combats.

Oblt. Hans Bethge shoots down a Sopwith single-seater at 19:05 over our territory south of Wingles. The falling machine rams a second Sopwith single-seater, which likewise falls apart and crashes near Wingles. Downed aircraft are confirmed by the Kogenluft as his 12th and 13th aerial victories.
According to available records the machines shot down were Sopwith Camels B3877and B3757 from 8 Squadron RNAS. Pilots F/S/Lt. E. A. Bennetts and F/Cdr. P. A. Johnston died.

18.08.17 Uffz. Josef Heiligers shoots down a Bristol Fighter at 07:55 over our territory about 1 km. south of Laillaing. Victory was contested by Lt. Viktor Schobinger from Jasta 12, which participated in the aerial combat, and it was granted to him.

20.08.17 Lt. Hans Schlieter is transferred via Armeeflugpark D to Fl. Abt. (A) 215.

21.08.17 Oblt. Hans Bethge shoots down a Martinsyde at 08:05 over our territory near Ennevelin. Downed aircraft is confirmed by the Kogenluft as his 14th aerial victory.
According to available records the downed machine was Martinsyde G.100 A6259 from 27 Squadron RFC. Pilot Capt. G. K. Smith, MC died.

22.08.17 Lt. Frhr. Rudolf von der Horst zu Hollwinkel is transferred from Jagdstaffelschule I to the Staffel.

Gefr. Josef Funk shoots down a Nieuport at 19:50 over our territory between Bauvin and Meurchin. Victory was at first contested by Uffz. Friedrich Gille from Jagdstaffel 12, which participated in the aerial combat, and Lt. Ernst Udet of Jagdstaffel 37, which likewise participated in the aerial combat, and through arbitration was granted to Gefr. Funk. Downed aircraft was confirmed by the Kogenluft as his 2nd aerial victory.
According to available records the downed machine was Nieuport B3473 from 40 Squadron RFC. Pilot Lt. H. A. Kennedy died.

24.08.17 From 17 to 23 August the Staffel undertook 145 operational flights with a total flight time of 155 flight hours, in the course of which it engaged in 28 aerial combats.

04.09.17 Oblt. Hans Bethge is appointed as leader of the Jagdgruppe of the 6. Armee consisting of Jagdstaffeln 12, 30, and 37.

Oblt. Hans Bethge shoots down a Bristol Fighter at 19:45 over our territory near Auchy. Five minutes later he shoots down a second Bristol Fighter over enemy territory to the west of Auchy. Downed aircraft are confirmed by the Kogenluft as his 15th and 16th aerial victories.

According to available records the downed machines were DH 4 A7480 and DH 4 A7487 from 25 Squadron RFC. The crew of the first machine, which crashed in German territory, 2/Lt. C. J. Pullen (pilot) and 2/Lt. E. D. S. Robinson (observer) died; of the crew of the latter machine, which crashed in Allied territory, observer 2/Lt. A. T. Williams was fatally wounded, while pilot Lt. C. A . Pike survived the crash uninjured.

09.09.17 Uffz. Emil Liebert is transferred from Jagdstaffelschule I to the Staffel.

17.09.17 Uffz. Emil Liebert shoots down a DH 5, which falls apart at 07:35 over our territory about 1 km. south of Izel. Victory was contested by Lt. Ernst Udet of Jagdstaffel 37, which participated in the aerial combat, and it was granted to him.

Uffz. Emil Liebert shoots down a DH 5 at 8:10 over our territory north of Vitry-en-Artois. Downed aircraft confirmed by the Kogenluft as his 1st aerial victory.

According to available records the downed machine was DH 5 A9409 from 41Squadron RFC. Pilot 2/Lt. R. E. Taylor died.

18.09.17 Oblt. Hans Bethge shoots down a Sopwith single-seater in flames at 10:35 over our territory northeast of Fontaine. Downed aircraft confirmed by the Kogenluft as his 17th aerial victory.
According to available records the downed machine was DH 5 A9426 from 41Squadron RFC. Pilot Lt. H. F. McArdle died.

19.09.17 Oblt. Hans Bethge shoots down a Morane monoplane at 12:03 over our territory near Le Cateau. Victory was contested by Oblt. Richard Flashar of Jagdstaffel 5, which participated in the aerial combat, and it was granted to him.

21.09.17 Vzfw. Max Willmann is transferred to Armeeflugpark 6.

Lt. Paul Erbguth shoots down a Sopwith single-seater at 09:15 over our territory near Monchy-le-Preux. Victory was contested by Gefr. Ulrich Neckel of Jagdstaffel 12, which participated in the aerial combat, and it was granted to him.

Lt. Paul Erbguth forces a Sopwith single-seater to land at 11:50 over our territory near Wevelghem. Downed aircraft confirmed by the Kogenluft as his 2nd aerial victory .
According to available records the downed machine was Sopwith Camel B3914 from 45 Squadron RFC. Pilot 2/Lt. E. A. Cooke fell into captivity unwounded.

24.09.17 Uffz. Otto Busch is transferred from Armeeflugpark 6 to the Staffel.

02.10.17 Vzfw. Hans Oberländer shoots down a Bristol Fighter at 18:10 over our territory about 2 km. northwest of Marquette. Downed aircraft confirmed by the Kogenluft as his 3rd victory.
According to available records the downed machine was Bristol F2b A7138 from 11 Squadron RFC. The crew of 2/Lt. J. M. McKenna (pilot) and 2/Lt. S. Sutcliffe (machine-gunner) died.

03.10.17 Lt. Hans von Schell is transferred via Armeeflugpark 6 to Jagdstaffel 37 as adjutant.

06.10.17 Lt. Hans Holthusen is transferred from Jagdstaffelschule I to the Staffel.
Uffz. Josef Heiligers is promoted to Vizefeldwebel.

Gefr. Josef Funk forced a Spad to land over our territory southeast of Seclin at 10:30. Downed aircraft is confirmed by the Kogenluft as his 3rd aerial victory.
According to available records the downed machine was Spad VII B3508 "C" from 19 Squadron RFC. Pilot 2/Lt. G. R. Long fell into captivity wounded.

11.10.17 In the afternoon a Sopwith single-seater landed at Phalempin airfield. Its pilot, 2/Lt. W. H. Winter of 28 Squadron RFC, who thought he was at an English airfield, was taken prisoner by the adjutant, Lt.

Douglas Schnorr; his machine, Sopwith Camel B6314, was captured intact.

12.10.17 Gefr. Josef Funk is promoted to Unteroffizier.

17.10.17 Sergt. Wilhelm Foege is transferred from Jagdstaffelschule I to the Staffel.

22.10.17 Oblt. Kurt Preissler is transferred from Jagdstaffelschule I to the Staffel.

27.10.17 Lt. Otto Fuchs shoots down a Sopwith single-seater in flames at 10:45 over our territory near the Harnes railway station. Downed aircraft is confirmed by the Kogenluft as his 2nd aerial victory.
According to available records the downed machine was Sopwith Camel B6374 from 43 Squadron RFC. Pilot 2/Lt. G. P. Bradley died.

29.10.17 Lt. Otto Fuchs shoots down a Sopwith over enemy territory near Gavrelle at 11:05. Downed aircraft confirmed by the Kogenluft as his 3rd victory.
According to available records the downed machine was most likely Sopwith Camel B2357 from 43 Squadron RFC. Pilot 2/Lt. C. H. Harriman died.

31.10.17 Oblt. Hans Bethge shoots down an R.E.8 at 13:10, which falls apart over our territory between Biache-St. Vaast and Plouvain. Downed aircraft confirmed by the Kogenluft as his 18th victory.
According to available records the downed machine was R.E.8 A3827 from 13 Squadron RFC. The crew of Lt. W. L. O. Parker (pilot) and 1/ AM H. L. Postons (machine-gunner) died.

01.11.17 Lt. Hans Schlieter, previously of Fl. Abt. (A) 215, arrives again at the Staffel.

02.11.17 The Jagdgruppe of the 6. Armee is disbanded.

08.11.17 Lt. Otto Fuchs is transferred to Jagdstaffel 77 via Bavarian FEA 1 in Schleissheim.

Lt. Hans Schlieter is promoted to Oberleutnant.

Vzfw. Hans Oberländer forces a DH 4 to land on our territory near the mine northwest of Monchecourt at 13:00. Downed aircraft confirmed by the Kogenluft as his 4th aerial victory.
According to available records the machine was DH 4 A7517 from 18 Squadron RFC. The crew of 2/Lt. W. C. Pruden (pilot) and 2/AM J. Conlin (machine gunner) fell into captivity unwounded.

09.11.17 From 2 to 8 November the Staffel undertook 26 operational flights in the course of which it engaged in 3 aerial combats.

10.11.17 Oblt. Bethge departs for four weeks of leave; Oblt. Kurt Preissler assumes acting leadership of the Staffel.

16.11.17 From 9 to 15 September the Staffel undertook 76 operational flights, in the course of which it engaged in 13 aerial combats.

20.11.17 The Staffel flies in the sector of the neighboring 2. Armee.

Vzfw. Josef Heiligers crashes in flames fatally during a ferrying flight in Albatros D V D.2196/17 at 13:47 over the railway triangle of Ostricourt.

22.11.17 Jagdstaffeln 18 and 30 form Jagdgruppe "North" of the 6. Armee. The leadership of the Jagdgruppe is assigned to the commander of Jagdstaffel 30.

23.11.17 From 16 to 22 November the Staffel undertook a total of 73 operational flights, in the course of which it engaged in 2 aerial combats.

02.12.17 Lt. Erich Kaus is transferred from Armeeflugpark 6 to the Staffel.

08.12.17 Uffz. Otto Busch is transferred to Armeeflugpark 6.

10.12.17 Oblt. Hans Bethge returns from leave and again assumes the leadership of the Staffel.

12.12.17 Jagdgruppe "North" of the 6. Armee now consists of Jagdstaffeln 18, 29, and 30.

14.12.17 Lt. Seewald is transferred from Jagdstaffelschule I to the Staffel.

15.12.17 Lt. Karl Weltz is transferred to the bombing instruction detachment at Frankfurt an der Oder.

21.12.17 Uffz. Otto Busch transferred back to the Staffel.

22.12.17 Sergt. Wilhelm Foege most probably flying Albatros D V D.4514/17 is shot down in aerial combat over Armentières and makes an emergency landing while severely wounded in the lung. After being recovered he is immediately brought to Feldlazarett 12, where he dies on the same day.
According to available records the opponents were apparently 7 Spads of 19 Squadron who reported the concerted downing of an enemy in an aerial combat with 8 Albatros D Vs at 14:20 south of Quesnoy.

28.12.17 Lt. Paul Erbguth is transferred as Staffelführer to the newly forming Jagdstaffel 54.

30.12.17 Lt. Reinhold Maier is transferred from Jagdstaffelschule I to the Staffel.

03.01.18 Vzfw. Hans Oberländer forces a Sopwith Camel to land on our territory near Billy at 14:50. The victory was first contested by Lt. Erich Kaus and through arbitration was granted to Vzfw. Oberländer. Downed aircraft confirmed by the Kogenluft as his 5th aerial victory.
According to available records the downed machine was Sopwith Camel N6351 from 10 Squadron RNAS. Pilot F/S/Lt. A. G. Beattie fell into captivity unwounded.

Uffz. Emil Liebert most probably flying Albatros D V D.2203/17 shoots down a Sopwith Camel at 14:50 over our territory near the Meurchin railway station. Downed aircraft confirmed by the Kogenluft as his 2nd aerial victory.
According to available records the downed machine was Sopwith Camel B5658 from 10 Squadron RNAS. Pilot F/S/Lt. F. Booth died.

Uffz. Emil Liebert lands close to his downed opponent at about 15:00 and while taking off once more collides at 16:05 with the guy-line of a captive balloon of Ballonzug 223, which had risen from Captive Balloon Field 23B, whereupon he crashes fatally. The torn loose balloon slowly drifts in the direction of Carvin and finally lands near Wingles.

10.01.18 Lt. Kurt Katzenstein is transferred via Armeeflugpark F to Jagdstaffel 55.

15.01.18 Oblt. Hans Bethge departs on leave for two weeks; Oblt. Kurt Preissler assumes acting leadership of the Staffel.

20.01.18 Lt. Seewald is transferred via FEA 6 in Grossenhain to Jagdstaffel 54.

22.01.18 Lt. Hans-Georg von der Marwitz shoots down a Sopwith single-seater at 13:55 over our territory between Bousbecque and Coucou. The victory was contested by Vzfw. Otto Fruhner of Jagdstaffel 26, which participated in the aerial combat, and it was granted to him.

23.01.18 Lt. Morhau is transferred from Jagdstaffelschule I to the Staffel.

26.01.18 Lt. Wendel Bastgen is transferred from Jagdstaffelschule 2 to the Staffel.

28.01.18 Lt. Reinhold Maier is wounded in aerial combat and is sent to a hospital. Further details are not known.

01.02.18 Jagdgruppe "North" of the 6. Armee now consists of Jagdstaffeln 29, 30, and 52.

06.02.18 Oblt. Hans Schlieter is transferred as Staffel commander to the newly forming Jagdstaffel 70.

09.02.18 Oblt. Hans Bethge returns from leave and a following detachment to Berlin-Adlershof and again assumes leadership of the Staffel.

10.02.18 Lt. Morhau is sent to hospital with an illness.

15.02.18 Oblt. Hans-Joachim Buddecke arrives at the Staffel for an introduction to the front.
Lt. Douglas Schnorr is transferred via Jagdstaffelschule I to Jagdstaffel 38 as adjutant.
Oblt. Kurt Preissler assumes the duties of adjutant.
Uffz. Otto Busch is transferred to Armee-Flugpark 6

16.02.18 Lt. Wendel Bastgen is shot down in Albatros D V D.4422/17 in aerial combat with S.E.5as over enemy territory southeast of Bailleul and after an emergency landing falls into British captivity.
According to available records the opponent was 2/Lt. P. J. Clayson from 1 Squadron RFC in S.E.5a B4881.

19.02.18 Oblt. Hans-Joachim Buddecke shoots down a Sopwith Camel at 14:00 over enemy territory between La Bassée and Neuve Chapelle. Downed aircraft confirmed by the Kogenluft as his 13th aerial victory.
According to available records the downed machine was a Sopwith Camel, serial number unknown, from 80 Squadron RFC which made an emergency landing in the forward Allied lines. Pilot Lt. S. L. H. Potter was wounded in the preceding aerial combat and went to a hospital.

Oblt. Hans Bethge forces a Sopwith single-seater to land at 14:05 on our territory between Aubers and Holpegarde. Downed aircraft confirmed by the Kogenluft as his 20th victory.
According to available records the downed machine was Sopwith Camel B9171 from 80 Squadron RFC. Pilot 2/Lt. E. Westmoreland fell into captivity wounded.

Lt. Hans-Georg von der Marwitz shoots down a Sopwith single-seater in flames at 14:05 over our territory near Marquillies. Downed aircraft confirmed by the Kogenluft as his 3rd victory.
According to available records the downed machine was Sopwith Camel B9185 from 80 Squadron RFC. Pilot 2/Lt. S. R. Pinder died.

08.03.18 Oblt. Hans-Joachim Buddecke is ordered to Jagdstaffel 18 as acting Staffelführer.

10.03.18 Oblt. Hans Bethge shoots down a DH 4 at 12:10 over our territory near Allennes, which bursts into flames upon impact. Downed aircraft confirmed by the Kogenluft as his 19th victory.
According to available records the downed machine was DH 4 A7719 from 18 Squadron RFC. The crew of 2/Lt. J. N. B. McKim (pilot) and Lt. C. R. H. Ffolliott (observer) died.

16.03.18 Lt. Hans-Georg von der Marwitz shoots down an English two-seater at 14:20 over enemy territory near Richebourg. Downed aircraft confirmed by the Kogenluft as his 4th victory.

17.03.18 Oblt. Hans Bethge, while attacking an English Sopwith formation in Pfalz D IIIa D.5888/17, is fatally wounded by return fire and crashes near Passchendaele around 12:55.
According to available records, his opponents were Capt. Alexander Roulstone, MC (pilot) and 2/Lt. William C. Venmore (observer) in DH 4 A7901 from 57 Squadron RFC, both of whom were wounded in the encounter.

For the time being Lt. Hans-Georg von der Marwitz assumes acting leadership of the Staffel.

19.03.18 Oblt. Kurt Preissler is appointed as the new Staffelführer.
Uffz. Max Marczinke is transferred from Jagdstaffelschule II to the Staffel.

20.03.18 Transport of the corpse of Oblt. Hans Bethge to Berlin. Staffelführer Oblt. Kurt Preissler is ordered to Berlin to participate in the burial ceremonies for Oblt. Hans-Joachim Buddecke and Oblt. Hans Bethge; Lt. Hans-Georg von der Marwitz assumes acting leadership of the Staffel.

30.03.18 Oblt. Harald Auffarth, Staffelführer of Jagdstaffel 29, is ordered to the Staffel to strengthen the Staffel leadership and arrives at the Staffel.

Uffz. Max Marczinke is shot down in Pfalz D IIIa D.8078/17 by anti-aircraft fire over Ploegsteert Wood and forced to land in enemy territory. Fell into English captivity.

01.04.18 Staffelführer Oblt. Kurt Preissler returns from Berlin and again assumes leadership of the Staffel.

07.04.18 Vzfw. Wilhelm Wagner transferred from Armeeflugpark 18 to the Staffel.

12.04.18 Lt. Hans-Georg von der Marwitz forces an English single-seater to land at 08:30 in our territory near Wambrechies. Downed aircraft confirmed by the Kogenluft as his 7th aerial victory.
According to available records the downed machine was Sopwith Camel B5424 from 54 Squadron RFC. Pilot 1/Lt. J. R. Sanford fell into captivity wounded.

Lt. Hans-Georg von der Marwitz shoots down an English single-seater in flames at 11:25 over our territory near Aubers. Downed aircraft confirmed by the Kogenluft as his 8th victory.
According to available records the downed machine was Sopwith Camel D1850 from 73 Squadron RFC. Pilot 2/Lt. M. F. Korslund died.

16.04.18 Oblt. Kurt Preissler is transferred to Jagdgeschwader II.
Lt. Hans-Georg von der Marwitz is appointed as the new Staffelführer.

19.04.18 Vzfw. Arthur Schiebler is transferred from Jagdstaffelschule II to the Staffel.
Vzfw. Hans Oberländer is promoted to Leutnant.

26.04.18 Lt. Kurt Katzentstein is transferred to the Staffel again from FEA 2 in Schneidemühl.

05.05.18 Uffz. Paul Marszinski is transferred from Jagdstaffelschule I to the Staffel.

06.05.18 Lt. Reinhold Maier again arrives at the Staffel after recovery from his wounds.

Lt. Frhr. von der Horst, since he is no longer fit for flying duties, is placed at the disposal of Idflieg.

10.05.18 Oblt. Harald Auffarth returns again to Jagdstaffel 29.

11.05.18 Lt. Ewald Siempelkamp, previously with construction supervision at the Brandenburg-Werke, arrives at the Staffel as the new adjutant.

14.05.18 Lt. Hans Eggersh, previously on the staff of the Bavarian Jagdgruppe, is transferred to the Staffel.

16.05.18 Lt. Hans-Georg von der Marwitz forces a Sopwith Camel to land on our territory near Lorgies at 08:40. Downed aircraft confirmed by the Kogenluft as his 5th victory.
According to available records the downed machine was Sopwith Camel D9540 from 208 Squadron RAF. Pilot Lt. W. E. Cowan fell into captivity unwounded.

18.05.18 Lt. Hans Eggersh shoots down an S.E.5a at 11:45 over our territory near Lestrem. The victory is contested by Uffz. Karl Pech of Jagdstaffel 29, which participated in the aerial combat, and it was granted to him.

20.05.18 Lt. Hans Oberländer shoots down an S.E.5a at 11:00 over our territory between Vieille Chapelle and Locon. Downed aircraft confirmed by the Kogenluft as his 6th victory.
According to available records the downed machine was S.E.5a D3438 from 40 Squadron RAF. Pilot Lt. G. Watson died.

21.05.18 Lt. Hans Holthusen forces an S.E.5a to land in our territory between Calonne and Pacaut Wood at 11:00. Downed aircraft confirmed by the Kogenluft as his 1st aerial victory.

23.05.18 Lt. Hans Oberländer is badly wounded in the shoulder during an aerial combat with a D.H.9 and after a successful emergency landing is sent to Feldlazarett 74.

27.05.18 Vzfw. Arthur Schiebler is shot down in a Pfalz DIIIa at 21:10 in aerial combat with S.E.5s and crashed fatally near Douvrin. Machine caught fire upon impact.
According to available records the opponents were S.E.5as from 40 Squadron RAF, who reported the downing of three opponents in a combat with 8 Pfalz D IIIs, namely at 20:10 by Lt. I. F. Hind in S.E.5a B675,"crashed" east of Billy, at 20:15 by Maj. R. S. Dallas in S.E.5a D3520, "crashed" northeast of La Bassée, and at 20:25 by Capt. G. H. Lewis in S.E.5a D3540, in flames near Hulluch.

31.05.18 Lt. Erich Kaus lightly wounded in aerial combat and sent to the hospital.

09.06.18 Lt. Hans-Georg von der Marwitz forces a Sopwith to land at 09:11 on our territory near Vieux Bequin. Downed aircraft confirmed by the Kogenluft as his 6th aerial victory.
According to available records the downed machine was Sopwith Camel B7163 from 210 Squadron RAF. Pilot 2/Lt. C. Marsden fell into captivity unwounded.

Lt. Kurt Katzenstein shoots down an English single-seater in flames at 09:20 over our territory near La Gorgue. Downed aircraft confirmed by the Kogenluft as his 1st victory.
According to available records the downed machine was Sopwith Camel D3348 from 210 Squadron RAF. Pilot 2/Lt. W. Breckenridge fell into captivity badly wounded.

11.06.18 Lt. Hans Oberländer is transferred to FEA 14 in Halle.

17.06.18 Lt. Hans-Georg von der Marwitz is wounded in the upper thigh during aerial combat, supposedly by mistaken fire from a machine-gun on the ground and after a smooth landing at his own airfield is taken to a hospital. Lt. Hans Eggers takes over temporary leadership of the Staffel.

20.06.18 Uffz. Josef Funk is promoted to Vizefeldwebel.

21.06.18 Lt. Reinhold Maier shoots down a captive balloon in flames at 16:30 over enemy territory to the south of Béthune near Aix-Noulette. The downed balloon is confirmed by the Kogenluft as his 1st victory.

22.06.18 A ten-day period of rest is ordered for the Staffel.
The commander of Jagdgruppe 3, Oblt. Richard Flashar, is additionally put in charge of the leadership of the Staffel.

24.06.18 Vzfw. Wilhelm Wagner is awarded the pilot's badge.

28.06.18 Vzfw. Josef Funk is transferred to FEA 10 in Böblingen.

30.06.18 Lt. Otto Franke and Lt. August Hartmann are transferred from Jagdstaffelschule I to the Staffel.

01.07.18 The Staffel is again ready for deployment.
Lt. Kurt Müller, previously of Jagdstaffel 24, is appointed as the new Staffelführer.

08.07.18 Vzfw. Hermann Benzier, Vzfw. Ernst David, and Uffz. Ehrlich are transferred from Armeeflugpark 6 to the Staffel.

10.07.18 Lt. Hans Holthusen forces a Sopwith Camel to land on our territory northwest of Armentières between Kemmel and L'Epinette. The victory is at first contested by Lt. Guntrum of Jagdstaffel 52, which participated in the aerial combat, and through arbitration was granted to Lt. Holthusen. Downed aircraft confirmed by the Kogenluft as his 2nd aerial victory.
According to available records the downed machine was Sopwith Camel D9500 from 43 Squadron RAF. Pilot Lt. C. B. Ridley fell into captivity wounded.

Lt. August Hartmann is lightly wounded in aerial combat. Further details are not known.
According to available records the opponents were possibly 7 Sopwith Camels from 43 Squadron RAF, who reported the downing of a Pfalz DIII by Lt. R. E. Meredith in Sopwith Camel D1870 at 10:00 (thus in the above combat) near Laventie.

14.07.18 Lt. Hans Holthusen and Lt. Reinhold Maier each shoot down an S.E.5a at 09:40 over enemy territory near Merris. The victories were not confirmed.

15.07.18 Lt. Hans-Georg von der Marwitz arrives at the Staffel again after recuperating from his wound.

17.07.18 Lt. Otto Franke forces a Sopwith Dolphin to land on our territory near Erquinghem at 19:00. Downed aircraft confirmed by the Kogenluft as his 1st victory.
According to available records the downed machine was Sopwith Dolphin C3792 from 19 Squadron RAF. Pilot 2/Lt. R. E. White fell into captivity unwounded.
While circling over his downed opponent, Lt. Otto Franke is mistakenly fired upon by German machine-guns on the ground and crashes fatally at 19:10.

22.07.18 Uffz. Paul Marszinski is shot down in aerial combat with S.E.5as at 09:50 and crashes fatally near Pont Maudit.
According to available records the opponents were S.E.5as from 40 Squadron RAF, who reported the downing of two Fokker DVIIs near Carvin at 08:50; one each by 1/Lt. R. G. Landis in S.E.5a E1318 and Lt. G. J. Strange in S.E.5a D3527.

24.07.18 Lt. Reinhold Maier shoots down a captive balloon in flames at 22:00 over enemy territory northwest of Grénay. Downed balloon confirmed by the Kogenluft as his 2nd victory.

25.07.18 Pursuant to an order of the Kogenluft, Lt. Hans-Georg von der Marwitz is to be assigned again as Staffelführer.

30.07.18 Lt. Kurt Müller is transferred as Staffelführer to Jagdstaffel 72 and Lt. Hans-Georg von der Marwitz again becomes Staffelführer of the Staffel.

Lt. Hans-Georg von der Marwitz shoots down a Sopwith Camel at 12:30 over enemy territory near Nieppe Forest west of Merville. Downed aircraft confirmed by the Kogenluft as his 9th aerial victory.

01.08.18 Lt. Hans-Georg von der Marwitz departs for two weeks leave; Lt. Hans Holthusen assumes acting Staffelführer.

06.08.18 Jagdstaffeln 14, 29, 30, and 52 are combined as Jagdgruppe "South" of the 6. Armee.

07.08.18 Vzfw. Ernst David crashed fatally from an altitude of about 5000 m. near Fleurbaix as a result of his Roland D VIa falling apart in combat with 18 S.E.5as and Sopwith Dolphins.
Vzfw. Hermann Benzier is injured during an emergency landing after an aerial combat and is sent to a hospital. Further details are not known.

15.08.08 Lt. Hans-Georg von der Marwitz returns from leave and again assumes leadership of the Staffel.

16.08.18 Flg. Erich Bieler is transferred from Jagdstaffelschule I to the Staffel.

19.08.18 Vzfw. Alfred Jaeschke is transferred from Jagdstaffelschule I to the Staffel.

At 08:15 a large English bombing raid with approximately 40 aircraft takes place at Phalempin airfield, which however causes only minor damage to the hangars, since at the time of the raid almost all of the machines were in the air.

The Staffel nonetheless receives orders to evacuate Phalempin airfield and begin the move to Avelin near Seclin.

23.08.18 From the new airfield the Staffel flies support daily in the 17. Armee sector.

24.08.18 Lt. Friedrich Bieling is transferred from Jagdstaffelschule I to the Staffel.

27.08.18 Move to Avelin airfield is completed.

Lt. Hans-Georg von der Marwitz shoots down a Sopwith at 07:40 over enemy territory near Tilloy. Downed aircraft is presented as his 10th victory to the Kogenluft for a decision.

28.08.18 Lt. Hans-Georg von der Marwitz is promoted to Oberleutnant.

29.08.18 Oblt. Hans-Georg von der Marwitz shoots down a Sopwith Dolphin in flames at 08:10 over our territory near Haplincourt. Downed aircraft is presented as his 11th victory to the Kogenluft for a decision. *According to available records the downed machine was S.E.5a C9061 "4" from 64 Squadron RAF. Pilot Lt. E. A. Parnell died.*

Lt. Ewald Siempelkamp, the Staffel Offz.z.b.V., is sent to a hospital for the treatment of old wounds.

01.09.18 Lt. Reinhold Maier assumes the duties of Staffel adjutant.

05.09.18 End of support activities in the 17. Armee sector.

17.09.18 Lt. Hans Oberländer is transferred after recuperating from his wounds via FEA 14 in Halle back to the Staffel.

Flg. Philipp Steiner is transferred from Armeeflugpark 6 to the Staffel.

22.09.18 Oblt. Hans-Georg von der Marwitz shoots down a Sopwith Camel at 08:15 over enemy territory south of Neuve Eglise. Downed aircraft is presented as his 12th victory to the Kogenluft for a decision.

Lt. Friedrich Bieling shoots down a Sopwith Camel at 08:20 over enemy territory near Ploegsteert Wood. Downed aircraft is presented as his 1st victory to the Kogenluft for a decision.

27.09.18 Vzfw. Wilhelm Wagner is transferred to Armeeflugpark 6.

28.09.18 Jagdstaffeln 30, 43, 52, and 63 are combined as Jagdgruppe 6. Armee under the leadership of Oblt. Adolf Gutknecht, Staffelführer of Jagdstaffel 43.

29.09.18 Lt. Hans Holthusen is ordered to Jagdstaffel 29 as acting Staffelführer.

30.09.18 Staffel moved to Sin airfield.

01.10.18 Staffel moved to Baisieux airfield on the road Lille – Tournai.

08.10.18 Oblt. Hans-Georg von der Marwitz shoots down an S.E.5 at 15:15 over enemy territory west of Cambrai near Gouzeaucourt. Downed aircraft is presented as his 13th victory to the Kogenluft for a decision.

10.10.18 Staffel moved to Anwaing airfield on the road Leuze - Renaix.

11.10.18 Lt. Hans Eggersh is ordered to Bombengeschwader 4 as a Staffel commander.

19.10.18 Lt. Reinhold Maier "wipes out" in a turn during a practice flight and hits the ground near Ellignies. He is recovered with a severe fracture in the base of his skull and dies on the transport to the hospital.

21.10.18 Jagdgruppe 6. Armee with Jagdstaffeln 30, 43, 52, and 63 is strengthened by Jagdstaffeln 28 and 33 and renamed Jagdgruppe 7b. The new Jagdgruppe leader is the Staffelführer of Jagdstaffel 28, Emil Thuy.

23.10.18 Lt. Ewald Prüfer is transferred from Jagdstaffelschule II to the Staffel.

27.10.18 Lt. Hans-Georg von der Marwitz shoots down an enemy aircraft over enemy territory. Downed aircraft is presented as his 14th victory to the Kogenluft for a decision. Further details about this downed aircraft are not known.

28.10.18 Vzfw. Alfred Jaeschke has to make an emergency landing in enemy territory near Tournai and falls into English captivity. Further details are not known.

31.10.18 Staffel moved to Vollezeele airfield.

11.11.18 Armistice.

Right: Formal portrait of Oblt. Hans Bethge (A. Imrie)

Logbook of Lt. Hans Holthusen, Jagdstaffel 30

Date	Aircraft	Time	Weather	Description
06.10.17				Transferred from Jagdstaffelschule I in Valenciennes to the Staffel.
08.10.17	Alb. D III 799/17	Morning	Very gusty	Dur. 18 min., Max. Alt. 2400 m, Over Phalempin airfield.
09.10.17	Alb. D V 2199/17	Morning	Very gusty	Dur. 15 min., Max. Alt. 700 m, Over Phalempin airfield.
09.10.17	Alb. D V 2199/17	Afternoon	Clear	Dur. 8 min., Max. Alt. 1200 m, Flight broken off.
09.10.17	Alb. D V 2199/17	Afternoon	Clear	Dur. 25 min., Max. Alt. 2800 m, Over Phalempin airfield.
10.10.17	Alb. D V 2199/17	Afternoon	Clear	Dur. 35 min., Max. Alt. 3800 m, Flight Path 6th Army., flak fire.
11.10.17	Alb. D V 2199/17	Morning	Clouds	Dur. 65 min., Max. Alt. 4000 m, Flight Path 6th Army. flak fire.
12.10.17	Alb. D V 2199/17	Afternoon	Rain	Dur. 25 min., Max. Alt. 1000 m, Flight Path 6th Army.
14.10.17	Alb. D V 2199/17	Morning	Mist	Dur. 15 min., Max. Alt. 2000 m, Flight broken off.
14.10.17	Alb. D V 2199/17	Midday	Mist	Dur. 20 min., Max. Alt. 4000 m, Flight Path 6th Army.
14.10.17	Alb. D V 2199/17	Afternoon	Mist	Dur. 60 min., Max. Alt. 4000 m, Flight Path 6th Army., flak fire.
15.10.17	Alb. D V 2199/17	Morning	Mist	Dur. 45 min., Max. Alt. 4300 m, Flight Path 6th Army.
15.10.17	Alb. D V 2199/17	Morning	Mist	Dur. 45 min., Max. Alt. 3800 m, Flight Path 6th Army.
16.10.17	Alb. D V 2126/17	Morning	Fair	Dur. 90 min., Max. Alt. 5200 m, Flight Path 6th Army., flak fire.
17.10.17	Alb. D V 2126/17	Morning	Fair	Dur. 80 min., Max. Alt. 5100 m, Flight Path 6th Army., flak fire.
17.10.17	Alb. D V 2126/17	Morning	Fair	Dur. 70 min., Max. Alt. 5000 m, Flight Path 6th Army., flak fire.
17.10.17	Alb. D V 2126/17	Afternoon	Overcast	Dur. 25 min., Max. Alt. 2100 m, Flight broken off.
18.10.17	Alb. D V 2126/17	Morning	Overcast	Dur. 65 min., Max. Alt. 4800 m, Flight Path 4th Army., 1 aerial combat with 9 Sopwiths on other side near Zonnebeke, flak fire.
18.10.17	Alb. D V 2126/17	Afternoon	Partially overcast	Dur. 90 min., Max. Alt. 4500 m, Flight Path 6th Army., flak fire.
20.10.17	Alb. D V 2126/17	Morning	Misty	Dur. 70 min., Max. Alt. 5000 m, Flight Path 6th Army.
20.10.17	Alb. D V 2126/17	Afternoon	Misty	Dur. 60 min., Max. Alt. 4500 m, Flight Path 6th Army.
20.10.17	Alb. D V 2126/17	Afternoon	Very misty	Dur. 50 min., Max. Alt. 3400 m, Flight Path 6th Army., flak fire.
21.10.17	Alb. D V 2126/17	Morning	Misty	Dur. 100 min., Max. Alt. 5300 m, Flight Path 4th Army., 1 aerial combat 4 Sopwiths near Armentières, flak fire.

Date	Aircraft	Time	Weather	Description
21.10.17	Alb. D V 2126/17	Afternoon	Misty	Dur. 70 min., Max. Alt. 5000 m, Flight Path 6th Army., 1 Aerial combat with 2 R.E.8's and 4 Sopwiths, flak fire.
21.10.17	Alb. D V 2126/17	Afternoon	Misty	Dur. 50 min., Max. Alt. 3000 m, Flight Path 6th Army.
22.10.17	Alb. D V 2126/17	Morning	Overcast	Dur. 75 min., Max. Alt. 4500 m, Flight Path 6th Army.
24.10.17	Alb. D V 2126/17	Morning	Overcast, gusty	Dur. 70 min., Max. Alt. 5000 m, Flight Path 6th Army.
24.10.17	Alb. D V 2126/17	Afternoon	Overcast, gusty	Dur. 80 min., Max. Alt. 5100 m, Flight Path 6th Army., flak fire.
24.10.17	Alb. D V 2126/17	Afternoon	Overcast, gusty	Dur. 50 min., Max. Alt. 1600 m, Flight Path 6th Army., flak fire.
27.10.17	Alb. D V 2126/17	Morning	Overcast	Dur. 60 min., Max. Alt. 3800m, Flight Path 6th Army., flak fire.
27.10.17	Alb. D V 2126/17	Afternoon	Overcast	Dur. 70 min., Max. Alt. 4800 m, Flight Path 6th Army., flak fire.
27.10.17	Alb. D V 2126/17	Afternoon	Overcast	Dur. 40 min., Max. Alt. 3000 m, Flight Path 6th Army., flak fire.
28.10.17	Alb. D V 2126/17	Morning	Overcast	Dur. 30 min., Max. Alt. 4000 m, Flight Path 6th Army., heavy flak fire.
29.10.17	Alb. D V 2126/17	Morning	Overcast	Dur. 70 min., Max. Alt. 4600 m, Flight Path 6th Army., flak fire.
31.10.17	Alb. D V 2126/17	Morning	Overcast	Dur. 90 min., Max. Alt. 4300 m, Flight Path 6th Army., 1 Aerial combat with 4 Sopwiths near, 1 Aerial combat with 3 B.F.s near La Bassée, flak fire.
31.10.17	Alb. D V 2126/17	Afternoon	Overcast	Dur. 80 min., Max. Alt. 4500 m, Flight Path 6th Army., flak fire.
07.11.17	Alb. D V 2126/17	Afternoon	Cloudy	Dur. 55 min., Max. Alt. 4000 m, Flight Path 6th Army., flak fire.
08.11.17	Alb. D V 2126/17	Morning	Cloudy	Dur. 60 min., Max. Alt. 4000 m, Flight Path 6th Army., 1 aerial combat with 6 B.F., jam, flak fire.
09.11.17	Alb. D V 2126/17	Morning	Cloudy	Dur. 70 min., Max. Alt. 4000 m, Flight Path 6th Army.
09.11.17	Alb. D V 2126/17	Afternoon	Cloudy, rain	Dur. 65 min., Max. Alt. 3500 m, Flight Path 6th Army., flak fire.
10.11.17	Alb. D V 2126/17	Morning	Cloudy, rain	Dur. 30 min., Max. Alt. 1600 m, Flight Path 6th Army., flak fire.
11.11.17	Alb. D V 2126/17	Afternoon	Cloudy	Dur. 70 min., Max. Alt. 4400 m, Flight Path 6th Army., flak fire.
12.11.17	Alb. D V 2126/17	Afternoon	Heavy mist	Dur. 75 min., Max. Alt. 4300 m, Flight Path 6th Army.
13.11.17	Alb. D V 2126/17	Afternoon	Heavy mist	Dur. 60 min., Max. Alt. 4200 m, Flight Path 6th Army.
15.11.17	Alb. D V 2126/17	Afternoon	Cloudy	Dur. 80 min., Max. Alt. 4000 m, Flight Path 6th Army., flak fire.
15.11.17	Alb. D V 2126/17	Afternoon	Cloudy	Dur. 45 min., Max. Alt. 2800 m, Flight Path 6th Army., flak fire.

Date	Aircraft	Time	Weather	Description
16.11.17	Alb. D V 2126/17	Morning	Cloudy	Dur. 70 min., Max. Alt. 3000 m, Flight Path 6th Army., flak fire.
19.11.17	Alb. D V 2126/17	Morning	Cloudy	Dur. 60 min., Max. Alt. 800 m, Flight Path 6th Army., fired upon trenches near Hulluch, flak fire.
19.11.17	Alb. D V 2126/17	Afternoon	Cloudy	Dur. 65 min., Max. Alt. 900 m, Flight Path 6th Army.
20.11.17	Alb. D V 4514/17	Afternoon	Gusty	Dur. 25 min., Max. Alt. 2000 m, Flight Path 6th Army.
22.11.17	Alb. D V 4514/17	Afternoon	Cloudy	Dur. 70 min., Max. Alt. 1500 m, Flight Path 2nd Army., flak fire. Return flight after intermediary landing at Afla 240 at Abscon.
23.11.17	Alb. D V 4514/17	Morning	Cloudy	Dur. 70 min., Max. Alt. 1800 m, Flight Path 6th Army., flak fire.
23.11.17	Alb. D V 4514/17	Morning	Cloudy	Dur. 45 min., Max. Alt. 3000 m, Flight Path 6th Army., 1 aerial combat with 2 B.F. near Vitry.
02.12.17	Alb. D III 791/17	Morning	Stormy	Dur. 65 min., Max. Alt. 4400 m, Flight Path 2nd Army.
02.12.17	Alb. D III 791/17	Afternoon	Stormy	Dur. 35 min., Max. Alt. 2200 m, Flight Path 2nd and 6th Army. Ground loop due to side wind with subsequent wreck.
03.12.17	Alb. D V 2051/17	Morning	Stormy	Dur. 80 min., Max. Alt. 3800 m, Flight Path 6th Army., flak fire.
03.12.17	Alb. D V 2051/17	Morning	Stormy	Dur. 60 min., Max. Alt. 2400 m, Flight Path 6th Army.
05.12.17	Alb. D V 2051/17	Morning	Clear	Dur. 90 min., Max. Alt. 4300 m, Flight Path 2nd and 6th Army. 1 aerial combat with 7 B.F.s near Cambrai, flak fire.
05.12.17	Alb. D V 2051/17	Afternoon	Clear	Dur. 95 min., Max. Alt. 4000 m, Flight Path 2nd and 6th Army. 1 aerial combat with 8 Sopwiths near Vacquerie, flak fire.
07.12.17	Alb. D V 2051/17	Morning	Overcast	Dur. 60 min., Max. Alt. 1500 m, Flight Path 6th Army.
07.12.17	Alb. D V 2051/17	Afternoon	Overcast	Dur. 30 min., Max. Alt. 3000 m, Flight Path 6th Army., flak fire.
10.12.17	Alb. D V 2126/17	Morning	Clear	Dur. 65 min., Max. Alt. 3800 m, Flight Path 6th Army., flak fire.
10.12.17	Alb. D V 2126/17	Afternoon	Clear	Dur. 60 min., Max. Alt. 3500 m, Flight Path 6th Army.
12.12.17	Alb. D V 2126/17	Afternoon	Clear	Dur. 60 min., Max. Alt. 3800 m, Flight Path 6th Army., flak fire.
15.12.17	Alb. D V 2126/17	Morning	Clear	Dur. 70 min., Max. Alt. 4000 m, Flight Path 6th Army., flak fire.
15.12.17	Alb. D V 2126/17	Afternoon	Misty	Dur. 45 min., Max. Alt. 3500 m, Flight Path 6th Army.
18.12.17	Alb. D V 2126/17	Afternoon	Clear	Dur. 50 min., Max. Alt. 4000 m, Flight Path 6th Army.
22.12.17	Alb. D V 4514/17	Morning	Overcast	Dur. 60 min., Max. Alt. 2500 m, Flight Path 6th Army.

Date	Aircraft	Time	Weather	Description
28.12.17	Alb. D V 2126/17	Morning	Clear	Dur. 80 min., Max. Alt. 4200 m, Flight Path 6th Army, flak fire.
29.12.17	Alb. D V 2126/17	Afternoon	Clear	Dur. 70 min., Max. Alt. 4000 m, Flight Path 6th Army.
01.01.18	Alb. D V 2126/17	Afternoon	Clear	Dur. 70 min., Max. Alt. 4000 m, Flight Path 6th Army.
02.01.18	Alb. D V 2126/17	Morning	Clear	Dur. 60 min., Max. Alt. 3500 m, Flight Path 6th Army, flak fire.
03.01.18	Alb. D V 2126/17	Morning	Partially overcast	Dur. 65 min., Max. Alt. 4800 m, Flight Path 6th Army, flak fire.
04.01.18	Alb. D V 2126/17	Afternoon	Clear	Dur. 75 min., Max. Alt. 4500, Flight Path 6th Army, flak fire.
09.01.18	Alb. D V 2126/17	Morning	Overcast	Dur. 40 min., Max. Alt. 4500 m, Flight Path 6th Army.
13.01.18	Pfalz D IIIa 5891/17	Morning	Clear	Dur. 90 min., Max. Alt. 4200 m, Flight Path 6th Army, 1 aerial combat with 1 B.F., 1 aerial combat with 4 Sopwiths, auxiliary spar and aileron broken.
19.01.18	Alb. D V 4422/17	Morning	Stratus clouds	Dur. 65 min., Max. Alt. 4500 m, Flight Path 6th Army.
19.01.18	Alb. D V 2126/17	Afternoon	Stratus clouds	Dur. 70 min., Max. Alt. 5000 m, Flight Path 6th Army, flak fire.
20.01.18	Alb. D V 2126/17	Afternoon	Stratus clouds	Dur. min. 50, Max. Alt. 3000 m, Flight Path 6th Army.
22.01.18	Pfalz D IIIa 5891/17	Morning	Stratus clouds	Dur. 60 min., Max. Alt. 4800 m, Flight Path 6th Army.
22.01.18	Alb. D V 2126/17	Midday	Stratus clouds	Dur. 95 min., Max. Alt. 4200 m, Flight Path 6th Army, 1 aerial combat with 3 Sopwiths, flak fire.
22.01.18	Alb. D V 2126/17	Afternoon	Stratus clouds	Dur. 25 min., Max. Alt. 2000 m, Flight Path 6th Army, motor plating broke off.
24.01.18	Pfalz D IIIa 5891/17	Afternoon	Clear	Dur. 70 min., Max. Alt. 4800 m, Flight Path 6th Army 1 aerial combat with 1 D. H. 4.
25.01.18	Pfalz D IIIa 5891/17	Morning	Clear	Dur. 80 min., Max. Alt. 4700 m, Flight Path 6th Army, 1 aerial combat with 5 D. H. 4's, 1 aerial combat with 1 B.F.
28.01.18	Pfalz D IIIa 5891/17	Morning	Misty	Dur. 75 min., Max. Alt. 5400 m, Flight Path 6th Army, flak fire.
28.01.18	Pfalz D IIIa 5891/17	Afternoon	Misty	Dur. 50 min., Max. Alt. 4800 m, Flight Path 6th Army.
29.01.18	Pfalz D IIIa 5891/17	Afternoon	Misty	Dur. 65 min., Max. Alt. 4800 m, Flight Path 6th Army.
30.01.18	Pfalz D IIIa 5891/17	Morning	Misty	Dur. 75 min., Max. Alt. 5000 m, Flight Path 6th Army, 1 aerial combat with 5 B.F.s, jam.
03.02.18	Pfalz D IIIa 5891/17	Morning	Misty	Dur. 70 min., Max. Alt. 5200 m, Flight Path 6th Army, 1 aerial combat with 1 S.E.5.
04.02.18	Pfalz D IIIa 5891/17	Afternoon	Partially overcast	Dur. 55 min., Max. Alt. 3000 m, Flight Path 6th Army, flak fire.
05.02.18	Pfalz D IIIa 5891/17	Morning	Partially overcast	Dur. 65 min., Max. Alt. 4500 m, Flight Path 6th Army, flak fire.

Date	Aircraft	Time	Weather	Description
05.02.18	Pfalz D IIIa 5891/17	Afternoon	Partially overcast	Dur. 45 min., Max. Alt. 4300 m, Flight Path 6th Army.
06.02.18	D.F.W. C.V 6401/17	Morning	Low clouds	Oblt. Schliefer in observer's seat, flight to Flg. Abt. 7.
06.02.18	D.F.W. C.V 6401/17	Afternoon	Partially overcast sky	Continued flight to Flg. Abt. 18 and then back to Phalempin, entire flight duration 50 min., altitude between 200 and 1500 m.
16.02.18	Pfalz D IIIa 5891/17	Morning	Clear	Dur. 35 min., Max. Alt. 4300 m, Flight Path 6th Army, 1 aerial combat with 9 Sopwiths, cabane wire and auxiliary spar broken, flak fire.
16.02.18	Pfalz D IIIa 5888/17	Afternoon	Clear	Dur. 75 min., Max. Alt.4700 m, Flight Path 6th Army, flak fire.
17.02.18	Pfalz D IIIa 5888/17	Morning	Clear	Dur. 65 min., Max. Alt. 4500 m, Flight Path 6th Army.
17.02.18	Pfalz D IIIa 5888/17	Afternoon	Misty	Dur. 60 min., Max. Alt. 3500 m, Flight Path 6th Army.
18.02.18	Pfalz D IIIa 5891/17	Morning	Misty	Dur. 55 min., Max. Alt. 4500 m, Flight Path 6th Army.
19.02.18	Alb. D V 2126/17	Morning	Misty	Dur. 65 min., Max. Alt. 4200 m, Flight Path 6th Army.
19.02.18	Alb. D V 2126/17	Midday	Misty	Dur. 55 min., Max. Alt. 4000 m, Flight Path 6th Army, 1 aerial combat with 5 Sopwiths, jam.
21.02.18	D.F.W. C.V 6401/17	Morning	High clouds	Lt. Schnorr in observer's seat, flight to Jastaschule I.
21.02.18	D.F.W. C.V 6401/17	Afternoon	High clouds	Return flight to Phalempin, entire flight time 45 minutes.
26.02.18	D.F.W. C.V 6401/17	Morning	Clear	Two flights to Flg. Abt. 18 and back, the first with Oblt. von Boddien, the second with Lt. Schröder in the observer's seat, flight time each 25 minutes, max. alt. 1000, 800, 800, and 500 m.
06.03.18	Pfalz D IIIa 5891/17	Morning	Clear	Dur. 80 min., Max. Alt. 4600 m, Flight Path 6th Army, flak fire.
06.03.18	Pfalz D IIIa 5891/17	Morning	Clear	Dur. 90 min., Max. Alt. 4500 m, Flight Path 6th Army, 1 aerial combat with Sopwith and R. E. 8 near Bois du Biez, flak fire.
06.03.18	Pfalz D IIIa 5891/17	Afternoon	Clear	Dur. 100 min., Max. Alt. 4200 m, Flight Path 6th Army, flak fire.
07.03.18	Pfalz D IIIa 5891/17	Morning	Misty	Dur. 70 min., Max. Alt. 4600 m, Flight Path 6th Army, flak fire.
07.03.18	Pfalz D IIIa 5891/17	Afternoon	Very misty	Dur. 40 min., Max. Alt. 3500 m, Flight Path 6th Army, flak fire.
08.03.18	Pfalz D IIIa 5891/17	Morning	Misty	Dur. 85 min., Max. Alt. 5000 m, Flight Path 6th Army, 1 aerial combat with 5 S.E.5s near Douai, 1 aerial combat with 4 B.F.s near Bois du Biez.
09.03.18	Pfalz D IIIa 5891/17	Morning	Misty	Dur. 75 min., Max. Alt. 4600 m, Flight Path 6th Army, flak fire near Béthune.
09.03.18	Pfalz D IIIa 5891/17	Morning	Misty	Dur. 70 min., Max. Alt. 4500 m, Flight Path 6th Army.

Date	Aircraft	Time	Weather	Description
10.03.18	Pfalz D IIIa 5888/17	Morning	Misty	Dur. 80 min., Max. Alt. 4600 m, Flight Path 6th Army, 1 aerial combat with 6 B.F.s, Bethge shoots down his 20th.
10.03.18	Pfalz D IIIa 5888/17	Afternoon	Misty	Dur. 80 min., Max. Alt. 4600 m, Flight Path 6th Army, aerial combat with 1 two-seater.
10.03.18	Pfalz D IIIa 5888/17	Afternoon	Misty	Dur. 40 min., Max. Alt. 3500 m, Flight Path 6th Army.
12.03.18	Pfalz D IIIa 5885/17	Morning	Misty	Dur. 70 min., Max. Alt. 4500 m, Flight Path 6th Army, 1 aerial combat with 10 S.E.5s.
12.03.18	Pfalz D IIIa 5885/17	Afternoon	Misty	Dur. 45 min., Max. Alt. 3800 m, Flight Path 6th Army, flak fire.
13.03.18	Pfalz D IIIa 5886/17	Morning	Misty	Dur. 60 min., Max. Alt. 5000 m, Flight Path 6th Army, flak fire.
13.03.18	Pfalz D IIIa 5886/17	Midday	Misty	Dur. 70 min., Max. Alt. 5100 m, Flight Path 6th Army, 1 aerial combat with A.W.s, Spads, Sopwiths, etc..
16.03.18	Pfalz D IIIa 5886/17	Morning	Misty	Dur. 80 min., Max. Alt. 4700 m, Flight Path 6th Army, 1 aerial combat with 5 B.F.s.
16.03.18	Pfalz D IIIa 5886/17	Midday	Misty	Dur. 75 min., Max. Alt. 5000 m, Flight Path 6th Army, von der Marwitz shoots down his 4th.
16.03.18	Pfalz D IIIa 5886/17	Afternoon	Misty	Dur. 65 min., Max. Alt. 5000 m, Flight Path 6th Army.
17.03.18	Pfalz D IIIa 5886/17	Morning	Stratus clouds	Dur. 65 min., Max. Alt. 4800 m, Flight Path 6th Army.
18.03.18	Pfalz D IIIa 5891/17	Morning	Stratus clouds	Dur. 65 min., Max. Alt. 4500 m, Flight Path 6th Army, 1 aerial combat with 10 S.E.5s and Sopwiths.
21.03.18	Pfalz D IIIa 5886/17	Morning	Misty	Dur. 90 min., Max. Alt. 4800 m, Flight Path 17th Army, 1 aerial combat with 4 Sopwiths, 1 aerial combat with 8 B.F.s.
22.03.18	Pfalz D IIIa 5886/17	Midday	Misty	Dur. 90 min., Max. Alt. 4800 m, Flight Path 17th Army.
22.03.18	Pfalz D IIIa 5886/17	Afternoon	Mist	Dur. 50 min., Max. Alt. 4500 m, Flight Path 17th Army.
23.03.18	Pfalz D IIIa 5886/17	Morning	Misty	Dur. 70 min., Max. Alt. 5100 m, Flight Path 17th Army.
23.03.18	Pfalz D IIIa 5886/17	Afternoon	Misty	Dur. 70 min., Max. Alt. 4600 m, Flight Path 17th Army.
24.03.18	Pfalz D IIIa 5886/17	Morning	Misty	Dur. 55 min., Max. Alt. 4500 m, Flight Path 17th Army.
25.03.18	Pfalz D IIIa 5886/17	Morning	Misty	Dur. 90 min., Max. Alt. 4200 m, Flight Path 17th and 6th Army, flak fire.
26.03.18	Pfalz D IIIa 5886/17	Morning	Clear	Dur. 80 min., Max. Alt. 4500 m, Flight Path 17th and 6th Army.
26.03.18	Pfalz D IIIa 5886/17	Afternoon	Gusty	Dur. 40 min., Max. Alt. 3000 m, Flight Path 6th Army.
27.03.18	Pfalz D IIIa 5891/17	Morning	Clouds and mist	Dur. 70 min., Max. Alt. 1000 m, Flight Path 17th Army.

Date	Aircraft	Time	Weather	Description
27.03.18	Pfalz D IIIa 5891/17	Afternoon	Mist	Dur. 50 min., Max. Alt. 1200 m, Flight Path 17th and 6th Army, flak fire.
28.03.18	Pfalz D IIIa 5886/17	Morning	Mist	Dur. 60 min., Max. Alt. 900 m, Flight Path 17th and 6th Army.
28.03.18	Pfalz D IIIa 5886/17	Afternoon	Strong wind	Dur. 30 min., Max. Alt. 2000 m, Flight Path 17th and 6th Army.
30.03.18	Pfalz D IIIa 5886/17	Morning	Stratus clouds	Dur. 65 min., Max. Alt. 3200 m, Flight Path 17th and 6th Army, flak fire.
01.04.18	Pfalz D IIIa 5886/17	Morning	Stratus clouds	Dur. 55 min., Max. Alt. 4000 m, Flight Path 17th and 6th Army.
01.04.18	Pfalz D IIIa 5891/17	Afternoon	Cumulous clouds	Dur. 65 min., Max. Alt. 4200 m, Flight Path 17th and 6th Army.
02.04.18	Pfalz D IIIa 5886/17	Afternoon	Cumulous clouds	Dur. 45 min., Max. Alt. 4200 m, Flight Path 17th and 6th Army.
07.04.18	Pfalz D IIIa 5891/17	Morning	Cumulous clouds	Dur. 40 min., Max. Alt. 3500 m, Flight Path 17th and 6th Army. 1 aerial combat with 3 Sopwith Camels.
10.04.18	Pfalz D IIIa 5891/17	Afternoon	Dull, rainy	Dur. 30 min., Max. Alt. 250 m, Flight Path 6th Army, machine-gun fire from ground.
11.04.18	Pfalz D IIIa 5891/17	Afternoon	Stratus clouds	Dur. 85 min., Max. Alt. 3000 m, Flight Path 6th Army, 1 aerial combat with R.E.8, flak fire.
12.04.18	Pfalz D IIIa 5891/17	Morning	Clear	Dur. 50 min., Max. Alt. 3000 m, Flight Path 6th Army, 1 aerial combat with Dolphins, von der Marwitz shoots down his 5th opponent.
12.04.18	Pfalz D IIIa 5891/17	Morning	Clear	Dur. 45 min., Max. Alt. 3000 m, Flight Path 6th Army, 1 aerial combat with S.E.s and Dolphins, von der Marwitz shoots down his 6th opponent.
12.04.18	Pfalz D IIIa 5891/17	Midday	Clear	Dur. 70 min., Max. Alt. 3500 m, Flight Path 6th Army, 1 aerial combat with D.H.4s and S.E.5s.
13.04.18	Pfalz D IIIa 5891/17	Midday	Dull, rainy	Dur. 50 min., Max. Alt. 300 m, Flight Path 6th Army, landing at La Gorgue, return flight in the afternoon, Dur. 30 min., Alt. 300 m.
15.04.18	Pfalz D IIIa 5891/17		Low-hanging clouds, rainy	Dur. 55 min., Max. Alt. 300 m, Flight Path 6th Army, 1 aerial combat with R.E. at Nieppe Forest.
17.04.18	Pfalz D IIIa 5891/17		"Foul"	Dur. 75 min., Max. Alt. 300 m, Flight Path 6th Army.
17.04.18	Pfalz D IIIa 5891/17		"Foul"	Dur. 60 min., Max. Alt. 200 m, Flight Path 6th Army.
18.04.18	Pfalz D IIIa 5891/17		"Foul"	Dur. 55 min., Max. Alt. 100 m, Flight Path 6th Army.
18.04.18	Pfalz D IIIa 5891/17		Snowfall	Dur. 35 min., Alt. 100-2000 m, Flight Path 6th Army.
20.04.18	Halb. C.		Cumulous clouds	With Oblt. Preissler in observer's seat, Flight Path Afla 209 at Toulis to Marle, Jasta 63 at Balâtre near Roye, Jasta 15 at Ham and back, Dur. 50, 40, 10, and 45 min., Distances 100, 80, 20, and 100 km.

Date	Aircraft	Time	Weather	Description
21.04.18	Pfalz D IIIa 5891/17		Cirrus clouds	Dur. 90 min., Max. Alt. 4600 m, Flight Path 6th Army, 1 aerial combat with 1 D.H., vicinity of Armentières.
21.04.18	Pfalz D IIIa 5891/17		Cirrus clouds	Dur. 50 min., Max. Alt. 3800 m, Flight Path 6th Army.
22.04.18	Pfalz D IIIa 5891/17		Cirrus clouds	Dur. 60 min., Max. Alt. 3500 m, Flight Path 6th Army.
23.04.18	Pfalz D IIIa 5891/17		Cirrus clouds	Dur. 60 min., Max. Alt. 4500 m, Flight Path 6th Army, flak fire.
29.04.18	Pfalz D IIIa 5891/17		Mist	Dur. 55 min., Max. Alt. 4500 m, Flight Path 6th Army.
02.05.18	Pfalz D IIIa 5891/17		Mist	Dur. 85 min., Max. Alt. 5300 m, Flight Path 6th Army.
02.05.18	Pfalz D IIIa 5891/17		Mist	Dur. 75 min., Max. Alt. 5000 m, Flight Path 6th Army.
03.05.18	Pfalz D IIIa 5891/17	Morning	Mist	Dur. 90 min., Max. Alt. 5200 m, Flight Path 6th Army, 1 aerial combat with S.E.5.
03.05.18	Pfalz D IIIa 5891/17	Morning	Mist	Dur. 90 min., Max. Alt. 5300 m, Flight Path 6th Army, 1 aerial combat with Camel, 1 aerial combat with 1 D.H.4, flak fire.
04.05.18	Pfalz D IIIa 5891/17	Afternoon	Stratus clouds	Dur. 75 min., Max. Alt. 5200 m, Flight Path 6th Army, 1 aerial combat with 2 B.F.s.
04.05.18	Pfalz D IIIa 5891/17	Afternoon	Cirrus clouds	Dur. 80 min., Max. Alt. 5200 m, Flight Path 6th Army, flak fire.
06.05.18	Pfalz D IIIa 5891/17	Afternoon	Mist	Dur. min. 70, Max. Alt. 4500 m, Flight Path 6th Army, flak fire.
07.05.18	Pfalz D IIIa 5891/17	Afternoon	Mist	Dur. 40 min., Max. Alt. 3000 m, Flight Path 6th Army.
07.05.18	Pfalz D IIIa 5891/17	Afternoon	Mist	Dur. 25 min., Max. Alt. 2500 m, Flight Path 6th Army.
08.05.18	Pfalz D IIIa 5891/17	Morning	Mist	Dur. 60 min., Max. Alt. 4400 m, Flight Path 6th Army.
08.05.18	Pfalz D IIIa 5891/17	Morning	Mist	Dur. 70 min., Max. Alt. 4600 m, Flight Path 6th Army, 1 aerial combat with 1 A.W. near Loos.
08.05.18	Pfalz D IIIa 5891/17	Afternoon	Heavy mist	Dur. 70 min., Max. Alt.4500 m, Flight Path 6th Army.
09.05.18	Pfalz D IIIa 5891/17	Morning	Heavy mist	Dur. 60 min., Max. Alt. 4800 m, Flight Path 6th Army, 1 aerial combat with 3 Sopwiths, fired continuously with 1 machine-gun.
09.05.18	Pfalz D IIIa 5891/17	Afternoon	Heavy mist	Dur. 90 min., Max. Alt. 5100 m, Flight Path 6th Army, flak fire.
11.05.18	Pfalz D IIIa 5891/17	Afternoon	Heavy mist	Dur. 75 min., Max. Alt. 4600 m, Flight Path 6th Army, 1 aerial combat with 6 S.E.5s, 1 S.E. at 1000 m over Béthune, flak fire.
11.05.18	Pfalz D IIIa 5891/17	Afternoon	Very misty	Dur. 70 min., Max. Alt. 4000 m, Flight Path 6th Army, 1 aerial combat with 3 Dolphins.

Date	Aircraft	Time	Weather	Description
14.05.18	Pfalz D IIIa 5891/17	Afternoon	Heavy mist	Dur. 60 min., Max. Alt. 4500 m, Flight Path 6th Army, 1 aerial combat D.H.4s.
14.05.18	Pfalz D IIIa 5891/17	Afternoon	Very misty	Dur. 60 min., Max. Alt.4500 m, Flight Path 6th Army, 1 aerial combat with 2 S.E.s near Festubert forced down from 4500 to 2000 m, 5 Camels above, flak fire.
15.05.18	Pfalz D IIIa 5891/17	Afternoon	Very misty	Dur. 60 min., Max. Alt. 4000 m, Flight Path 6th Army, 1 aerial combat with 5 S.E.s over Nieppe Forest and Hazebrouck, flak fire.
16.05.18	Pfalz D IIIa 5891/17	Morning	Clear	Dur. 80 min., Max. Alt. 4700 m, Flight Path 6th Army, 1 aerial combat with 3 S.E.5s, flak fire.
16.05.18	Pfalz D IIIa 5891/17	Morning	Misty	Dur. 90 min., Max. Alt. 5200 m, Flight Path 6th Army, 1 aerial combat with 16 Camels, flak fire.
17.05.18	Pfalz D IIIa 5891/17	Morning	Misty	Dur. 65 min., Max. Alt. 4500 m, Flight Path 6th Army, 1 aerial combat with 5 S.E.5s, 1 aerial combat with 10 Camels, 1 Camel over Merville forced down to 500 m.
17.05.18	Pfalz D IIIa 5891/17	Morning	Misty	Dur. 60 min., Max. Alt.4000 m, Flight Path 6th Army, 1 aerial combat with mixed formation. By 5 Sopwiths through . . .
17.05.18	Pfalz D IIIa 5891/17	Afternoon	Very misty	Dur. 75 min., Max. Alt. 4600 m, Flight Path 6th Army, flak fire.
18.05.18	Pfalz D IIIa 5891/17	Morning	Very misty	Dur. 75 min., Max. Alt. 4600 m, Flight Path 6th Army.
18.05.18	Pfalz D IIIa 5891/17	Morning	Very misty	Dur. 75 min., Max. Alt. 4400 m, Flight Path 6th Army, 1 aerial combat with 10 B.F.s, 1 aerial combat with 12 S.E.s; forced down to 50 m by the S.E.s.
19.05.18	Pfalz D IIIa 5891/17	Morning	Very misty	Dur. 90 min., Max. Alt. 4500 m, Flight Path 6th Army.
19.05.18	Pfalz D IIIa 5891/17	Afternoon	Very misty	Dur. 60 min., Max. Alt. 3000 m, Flight Path 17th Army, flight with Jastas 32 and 29. 1 aerial combat with 20 S.E.s near Arras.
20.05.18	Pfalz D IIIa 5891/17	Morning	Very misty	Dur. 65 min., Max. Alt. 4700 m, Flight Path 6th Army, 1 aerial combat with 3 S.E.s, . . . exercised down to close above Béthune; Oberländer's victory, flak fire.
20.05.18	Pfalz D IIIa 5891/17	Afternoon	Very misty	Dur. 100 min., Max. Alt. 5200 m, Flight Path 6th Army, flak fire.
21.05.18	Pfalz D IIIa 5891/17	Morning	Very misty	Dur. 75 min., Max. Alt. 4000 m, Flight Path 6th Army, 1 aerial combat with 3 S.E.5s, 1st victory attained!
21.05.18	Pfalz D IIIa 5891/17	Afternoon	Very hazy	Dur. 80 min., Max. Alt.2500 m, Flight Path 6th Army, flak fire.
22.05.18	Pfalz D IIIa 5891/17	Morning	Very misty	Dur. 75 min., Max. Alt. 4500 m, Flight Path 6th Army.
22.05.18	Pfalz D IIIa 5891/17	Morning	Misty	Dur. 30 min., Max. Alt.3500 m, Flight Path 6th Army.
23.05.18	Pfalz D IIIa 5891/17	Morning	Clear	Dur. 90 min., Max. Alt.4800 m, Flight Path 6th Army, Oberländer wounded by 1 Camel.

Date	Aircraft	Time	Weather	Description
23.05.18	Pfalz D IIIa 5891/17	Morning	Clouding over	Dur. 35 min., Max. Alt. 3000 m, Flight Path 6th Army.
25.05.18	Pfalz D IIIa 5891/17	Morning	Misty	Dur. 85 min., Max. Alt. 4900 m, Flight Path 6th Army.
27.05.18	Pfalz D IIIa 5891/17	Morning	Clear	Dur. 60 min., Max. Alt. 4500 m, Flight Path 6th Army, 1 aerial combat with 6 S.E.s.
28.05.18	Pfalz D IIIa 5891/17	Morning	Clear	Dur. 75 min., Max. Alt. 4700 m, Flight Path 6th Army, 1 aerial combat with S.E.5s and Dolphins, flak fire.
28.05.18	Pfalz D IIIa 5891/17	Morning	Clear	Dur. 75 min., Max. Alt. 3200 m, Flight Path 6th Army.
29.05.18	Pfalz D IIIa 5891/17	Afternoon	Clear	Dur. 95 min., Max. Alt. 4300 m, Flight Path 6th Army, flak fire.
29.05.18	Pfalz D IIIa 5891/17	Afternoon	Clear	Dur. 70 min., Max. Alt. 4700 m, Flight Path 6th Army, 1 aerial combat with 8 D.H.4s.
30.05.18	Pfalz D IIIa 5891/17	Afternoon	Clear	Dur. 65 min., Max. Alt. 4000 m, Flight Path 6th Army.
30.05.18	Pfalz D IIIa 5891/17	Afternoon	Clear	Dur. 75 min., Max. Alt. 4500 m, Flight Path 6th Army, flak fire.
31.05.18	Pfalz D IIIa 5891/17	Morning	Clear	Dur. 55 min., Max. Alt. 4200 m, Flight Path 6th Army, 1 aerial combat with 6 Sopwith Dolphins.
31.05.18	Pfalz D IIIa 5891/17	Morning	Clear	Dur. 65 min., Max. Alt. 4200 m, Flight Path 6th Army, 1 aerial combat with 6 S.E.5s, 1 aerial combat with 1 A.W.
04.07.18	Fok. D VII 370/18	Afternoon	Clear	Dur. 75 min., Max. Alt. 5000 m, Flight Path 6th Army.
05.07.18	Fok. D VII 370/18	Morning	Overcast	Dur. 60 min., Max. Alt. 3000 m, Flight Path 6th Army, flak fire.
05.07.18	Fok. D VII 370/18	Afternoon	Overcast	Dur. 60 min., Max. Alt. 3000 m, Flight Path 6th Army, flak fire.
05.07.18	Fok. D VII 370/18	Afternoon	Overcast	Dur. 60 min., Max. Alt. 4000 m, Flight Path 6th Army, 1 aerial combat with 2 D.H.9s, flak fire.
06.07.18	Fok. D VII 370/18	Morning	Clear	Dur. 85 min., Max. Alt. 5500 m, Flight Path 6th Army, 1 aerial combat with 8 S.E.5s and Sopwiths.
07.07.18	Fok. D VII 370/18	Morning	Clear	Dur. 60 min., Max. Alt. 5400 m, Flight Path 6th Army, flak fire.
07.07.18	Fok. D VII 370/18	Afternoon	Clear	Dur. 50 min., Max. Alt. 4000 m, Flight Path 6th Army, 1 aerial combat with 5 Sopwith Camels and 2 B.F.s.
08.07.18	Fok. D VII 370/18	Morning	Overcast	Dur. 65 min., Max. Alt. 4000 m, Flight Path 6th Army, 1 aerial combat with 5 Sopwiths, 1 aerial combat with 5 S.E.5s, 1 aerial combat with 8 B.F.s, flak fire.
08.07.18	Fok. D VII 370/18	Afternoon	Clear	Dur. 75 min., Max. Alt. 4300 m, Flight Path 6th Army, flak fire.
08.07.18	Fok. D VII 370/18	Afternoon	Overcast	Dur. 60 min., Max. Alt. 4000 m, Flight Path 6th Army, 1 aerial combat with 2 B.F.s and 5 Sopwiths, flak fire.

Date	Aircraft	Time	Weather	Description
09.07.18	Fok. D VII 370/18	Afternoon	Overcast	Dur. 45 min., Max. Alt. 3500 m, Flight Path 6th Army, 1 aerial combat with 1 D.H.4, flak fire.
10.07.18	Fok. D VII 370/18	Morning	Overcast	Dur. 50 min., Max. Alt. 4300 m, Flight Path 6th Army, 1 aerial combat with 17 Sopwiths; I shoot down a Camel as my 2nd aerial victory, flak fire.
11.07.18	Fok. D VII 370/18	Morning	Clear	Dur. 55 min., Max. Alt. 4800 m, Flight Path 6th Army, 1 aerial combat with 3 S.E.5s; . . . (?) lever broken.
11.07.18	Fok. D VII 370/18	Morning	Clouding over	Dur. 70 min., Max. Alt. 5000 m, Flight Path 6th Army.
12.07.18	Fok. D VII 370/18	Afternoon	Cloudy & stormy	Dur. 60 min., Max. Alt. 4500 m, Flight Path 6th Army, flak fire.
13.07.18	Fok. D VII 370/18	Morning	Partially overcast	Dur. 75 min., Max. Alt. 4800 m, Flight Path 6th Army, flak fire.
13.07.18	Fok. D VII 370/18	Afternoon	Partially overcast	Dur. 80 min., Max. Alt. 4000 m, Flight Path 6th Army, 1 aerial combat with 1 D.H.4 and 5 S.E.5s, flak fire.
14.07.18	Fok. D VII 370/18	Morning	Partially overcast	Dur. 80 min., Max. Alt. 4000 m, Flight Path 6th Army, 1 aerial combat with 1 D.H.4, 1 aerial combat with ca. 20 S.E.s, Lt. Maier 1 S.E. shot down, became very . . . (illegible), flak fire.
15.07.18	Fok. D VII 370/18	Afternoon	Partially overcast	Dur. 45 min., Max. Alt. 3000 m, Flight Path 6th Army.
15.07.18	Fok. D VII 370/18	Afternoon	Partially overcast	Dur. 60 min., Max. Alt. 4800 m, Flight Path 6th Army.
16.07.18	Fok. D VII 370/18	Afternoon	Partially overcast	Dur. 55 min., Max. Alt. 3000 m, Flight Path 6th Army, flak fire.
17.07.18	Fok. D VII 370/18	Morning	Partially overcast	Dur. 70 min., Max. Alt. 4500 m, Flight Path 6th Army, 1 aerial combat with 1 D.H.
17.07.18	Fok. D VII 370/18	Afternoon	Partially overcast	Dur. 70 min., Max. Alt. 4800 m, Flight Path 6th Army, 1 aerial combat with mixed formation, Lt. Franke shoots down a Dolphin, then crashes fatally himself.
18.07.18	Fok. D VII 370/18	Morning	Partially overcast	Dur. 65 min., Max. Alt. 5000 m, Flight Path 6th Army, flak fire.
18.07.18	Fok. D VII 370/18	Midday	Clear	Dur. 60 min., Max. Alt. 4800 m, Flight Path 6th Army.
19.07.18	Fok. D VII 370/18	Morning	Clear	Dur. 70 min., Max. Alt. 4700 m, Flight Path 6th Army, 1 aerial combat with 10 D.H.s, jam.
19.07.18	Fok. D VII 370/18	Midday	Several cloud layers	Dur. 45 min., Max. Alt. 3800 m, Flight Path 6th Army.
20.07.18	Fok. D VII 370/18	Morning	Clear	Dur. 80 min., Max. Alt. 5500 m, Flight Path 6th Army, 1 aerial combat with 12 D.H.s, jam.
20.07.18	Fok. D VII 370/18	Afternoon	Overcast	Dur. 45 min., Max. Alt. 4000 m, Flight Path 6th Army.
21.07.18	Fok. D VII 370/18	Afternoon	Partially overcast	Dur. 30 min., Max. Alt. 3700 m, Flight Path 6th Army, flak fire.

Date	Aircraft	Time	Weather	Description
22.07.18	Fok. D VII 370/18	Morning	Partially overcast	Dur. 70 min., Max. Alt. 4500 m, Flight Path 6th Army, 1 aerial combat with 4 S.E.5s.
22.07.18	Fok. D VII 370/18	Afternoon	Partially overcast	Dur. 45 min., Max. Alt. 4300 m, Flight Path 6th Army, 1 aerial combat with 3 D.H.4s.
24.07.18	Fok. D VII 370/18	Afternoon	Cumulous clouds	Dur. 60 min., Max. Alt. 4200 m, Flight Path 6th Army.
24.07.18	Fok. D VII 370/18	Afternoon	Cumulous clouds	Dur. 75 min., Max. Alt. 5000 m, Flight Path 6th Army, 1 aerial combat with 1 D.H.9, 2 jams.
24.07.18	Fok. D VII 370/18	Afternoon	Misty	Dur. 60 min., Max. Alt. 4500 m, Flight Path 6th Army, Lt. Maier shoots a balloon down in flames.
25.07.18	Fok. D VII 370/18	Afternoon	Misty	Dur. 60 min., Max. Alt. 4500 m, Flight Path 6th Army.
26.07.18	Fok. D VII 370/18	Morning	Cloud layers, rain	Dur. 35 min., Max. Alt. 3000 m, Flight Path 6th Army.
27.07.18	Fok. D VII 370/18	Morning	Partially overcast	Dur. 55 min., Max. Alt. 4600 m, Flight Path 6th Army.
28.07.18	Fok. D VII 370/18	Morning	Partially overcast	Dur. 50 min., Max. Alt. 4400 m, Flight Path 6th Army, heavy flak fire.
28.07.18	Fok. D VII 541/18	Afternoon	Overcast	Two take-offs, each time flight duration 5 minutes, flight altitude 300m.
29.07.18	Fok. D VII 370/18	Morning	Ground mist	Dur. 50 min., Max. Alt. 5000 m, Flight Path 6th Army.
29.07.18	Fok. D VII 370/18	Afternoon	Ground mist	Dur. 35 min., Max. Alt. 3500 m, Flight Path 6th Army.
30.07.18	Fok. D VII 370/18	Morning	Ground mist	Dur. 90 min., Max. Alt. 5000 m, Flight Path 6th Army, 1 aerial combat with10 to 20 two-seaters, 1 aerial combat with 1 Camel, Marwitz and I force the Camel down; von der Marwitz 9th victory.
31.07.18	Fok. D VII 370/18	Afternoon	Ground mist	Dur. 65 min., Max. Alt. 4500 m, Flight Path 6th Army, 1 aerial combat with 8 B.F.s and 15 single-seaters: 3 victories.
01.08.18	Fok. D VII 370/18	Morning	Mist	Dur. 80 min., Max. Alt. 5000 m, Flight Path 6th Army, 1 aerial combat with 5 Dolphins near Don.
01.08.18	Fok. D VII 370/18	Afternoon	Mist	Dur. 60 min., Max. Alt. 4500 m, Flight Path 6th Army, 1 aerial combat with 1 D.H.4, fired upon by German flak during the aerial combat.
01.08.18	Fok. D VII 370/18	Afternoon	Mist	Dur. 45 min., Max. Alt. 4500 m, Flight Path 6th Army.
03.08.18	Fok. D VII 370/18	Morning	Cloudy	Dur. 60 min., Max. Alt. 4800 m, Flight Path 6th Army, 1 aerial combat with S.E.5s and Sopwiths near Meteren.
03.08.18	Fok. D VII 370/18	Afternoon	Cloudy	Dur. 60 min., Max. Alt. 4800 m, Flight Path 6th Army, flak fire.
04.08.18	Fok. D VII 370/18	Morning	Partially overcast	Dur.55 min., Max. Alt. 4500 m, Flight Path 6th Army, flak fire.
04.08.18	Pfalz D.XII 1488/18	Morning	Partially overcast	Dur. 10 min., Max. Alt. 1000 m, Flight Path Park 6 – Jasta 30.

Date	Aircraft	Time	Weather	Description
04.08.18	Fok. D VII 370/18	Afternoon	Very misty	Dur. 60 min., Max. Alt. 5100 m, Flight Path 6th Army.
07.08.18	Fok. D VII 370/18	Morning	Lower cloud layer at 5100 m	Dur. 65 min., Max. Alt. 5000 m, Flight Path 6th Army, 1 aerial combat with 18 S.E.s and Sopwiths. Vfw. David shot down .
08.08.18	Fok. D VII 370/18	Morning	Cumulous clouds, mist	Dur. 50 min., Max. Alt. 2000 m, Flight Path Jasta 30 – 34 near Ennemain, 6th, 17th, 2nd Army.
08.08.18	Fok. D VII 370/18	Afternoon	Cloud ceiling at 800-1000 m	Dur. 50 min., Max. Alt. 1000 m, Flight Path Jasta 34 – Jasta 5 near Moislains, 2nd Army. Fired upon by our own flak.
08.08.18	Fok. D VII 370/18	Afternoon	Cloud ceiling at 800-1000 m	Dur. 70 min., Max. Alt. 1000 m, Flight Path Jasta 5 – Jasta 30, 2nd, 17th, 6th Army, including return flight.
09.08.18	Fok. D VII 370/18	Morning	Cloud ceiling at 800-1000 m	Dur. 30 min., Max. Alt. 2000 m, Flight Path Jasta 30 – Jasta 5, 6th, 17tn, and 2nd Army.
09.08.18	Fok. D VII 370/18	Morning	Clouds at 1500 m	Dur. 70 min., Max. Alt. 3000 m, Flight Path 2nd Army, flak fire.
09.08.18	Fok. D VII 370/18	Afternoon	Clouds at 1500 m	Dur. 50 min., Max. Alt. 2000 m, Flight Path 2nd Army.
09.08.18	Fok. D VII 370/18	Afternoon	Clouds at 1500 m	Dur. 40 min., Max. Alt. 2200 m, Flight Path 2nd Army, 1 aerial combat with 4 Sopwiths, flak fire.
09.08.18	Fok. D VII 370/18	Afternoon	Clouds at 1500 m	Dur. 45 min., Max. Alt. 1200 m, Flight Path Jasta 5 – Jasta 30, 2nd, 17th, 6th Army.
10.08.18	Fok. D VII 370/18	Morning	Clouds at 1500 m	Dur. 30 min., Max. Alt. 1500 m, Flight Path Jasta 30 – Jasta 5, 6th, 17th, 2nd Army.
10.08.18	Fok. D VII 370/18	Morning	Clouds at 1500 m	Dur. 60 min., Max. Alt. 4000 m, Flight Path 2nd Army.
10.08.18	Fok. D VII 370/18	Afternoon	Clouds at 1500 m	Dur. 60 min., Max. Alt. 3500 m, Flight Path 2nd Army.
10.08.18	Fok. D VII 370/18	Afternoon	Clouds at 1500 m	Return flight, see above.
11.08.18	Fok. D VII 370/18	Morning	Cumulous clouds	Dur. 50 min., Max. Alt. 4000 m, Flight Path 6th Army.
11.08.18	Fok. D VII 370/18	Afternoon	Mist	Dur. 50min., Max. Alt. 4000 m, Flight Path 6th Army, heavy flak fire.
12.08.18	Fok. D VII 370/18	Morning	Mist	Dur. 10 min., Max. Alt. 1800 m, flight broken off due to machine-gun defect.
13.08.18	Fok. D VII 370/18	Afternoon	Clear	Dur. 75 min., Max. Alt. 5000 m, Flight Path 6th Army.
14.08.18	Fok. D VII 370/18	Morning	Mist	Dur. 80 min., Max. Alt. 5000 m, Flight Path 6th Army.
14.08.18	Fok. D VII 370/18	Morning	Mist	Dur. 70 min., Max. Alt. 4500 m, Flight Path 6th Army.
14.08.18	Fok. D VII 370/18	Afternoon	Mist	Dur. 70 min., Max. Alt. 4000 m, Flight Path 6th Army.

Date	Aircraft	Time	Weather	Description
19.08.18	Fok. D VII 370/18	Afternoon	Clear, gusty	Dur. 40 min., Max. Alt. 2700 m, flight from Phalempin to the new airfield at Avelin.
22.08.18	Fok. D VII 370/18	Morning	Mist	Dur. 65 min., Max. Alt. 4200 m, Flight Path 6th Army, flak fire.
23.08.18	Fok. D VII 370/18	Morning	Mist	Dur. 60 min., Max. Alt. 4000 m, Flight Path 6th Army, flak fire.
23.08.18	Fok. D VII 370/18	Afternoon	Mist	Dur. 50 min., Max. Alt. 3000 m, Flight Path 6th Army, 1 aerial combat with 1 B.F., flak fire.
24.08.18	Fok. D VII 370/18	Afternoon	Mist	Dur. 65 min., Max. Alt. 5300 m, Flight Path 6th Army, flak fire.
25.08.18	Fok. D VII 370/18	Morning	Mist	Dur. 65 min., Max. Alt. 5000 m, Flight Path 6th Army, 1 aerial combat with 7 S.E.5s, flak fire.
25.08.18	Fok. D VII 370/18	Morning	Mist	Dur. 50 min., Max. Alt. 4300 m, Flight Path 6th Army, 1 aerial combat with 8 B.F.s, entire ammunition fired off, flak fire.
25.08.18	Fok. D VII 370/18	Afternoon	Mist	Dur. 55 min., Max. Alt. 4500 m, Flight Path 6th and 17th Army.
25.08.18	Fok. D VII 370/18	Afternoon	Mist	Dur. 40 min., Max. Alt. 4000 m, Flight Path 6th and 17th Army.
27.08.18	Fok. D VII 370/18	Morning	Mist	Dur. 75 2000 min., Max. Alt. m, Flight Path 6th and 17th Army, 1 aerial combat with 7 S.E.5s, flak fire.
16.09.18	Fok. D VII 370/18	Morning	Clear	Dur. 35 min., Max. Alt. 3500 m, Flight Path 6th Army.
16.09.18	Fok. D VII 370/18	Afternoon	Clear	Dur. 70 min., Max. Alt. 5000 m, Flight Path 6th Army, 1 aerial combat with mixed formation, flak fire.
20.09.18	Fok. D VII 370/18	Afternoon	Clear	Dur. 60 min., Max. Alt. 4000 m, Flight Path 6th Army, flak fire.
21.09.18	Fok. D VII 370/18	Afternoon	Clear	Dur. 75 min., Max. Alt. 4500 m, Flight Path 6th Army, 1 aerial combat with 1 B.F., flak fire.
22.09.18	Fok. D VII 370/18	Morning	Clear	Dur. 90 min., Max. Alt. 4500 m, Flight Path 6th Army, 1 aerial combat with 3 Camels, Lt. von der Marwitz and Lt. Bieling each shoot down a Camel, flak fire.
23.09.18	Fok. D VII 370/18	Afternoon	Clear	Dur. 60 min., Max. Alt. 4800 m, Flight Path 6th Army, flak fire.
24.09.18	Fok. D VII 370/18	Morning	Clear	Dur. 70 min., Max. Alt. 4500 m, Flight Path 6th Army, 1 aerial combat with 7 S.E.s, flak fire.
27.09.18	Fok. D VII 370/18	Morning	Clear	Dur. 60 min., Max. Alt. 5200 m, Flight Path 6th Army, flak fire. This is my 273rd and final pursuit flight with Jagdstaffel 30!
30.09.18				Transferred to Jagdstaffel 29 and appointed as commander of the Staffel.

Below: Excerpt of the logbook of Lt. Hans Holthusen from the 15th April 1918 to 24th April 1918. On the 21th April 1918 he achived his first victory in an air combat against three S.E.5a fighters.

Nr.	Datum	Führer	Beobachter	Flugzeug	Dauer	Höhe	[illegible]	Maschine	[illegible]
1021	15/4	selbst		6. Armee	55	300		Pfalz D3a 5986	[illegible]
1022	17/4	"		"	75	300		Pfalz D3a 5891	"
1023	"	"		"	60	200		"	"
1024	18/4	"		"	55	100		"	"
1025	19/4	"		"	35	10-2000		"	[illegible]
1026	20/4	"	[illegible]	[illegible]	50	1000	100	[illegible]	[illegible]
1027	"	"	— " —	[illegible]	70	1200	80	"	"
1028	"	"	— " —	[illegible]	10	500	20	"	"
1029	"	"		[illegible]	45	1000	100	"	"
1030	21/4	"		6. Armee	90	4600		Pfalz D3a 5891	[illegible]
1031	"	"		"	50	3800		"	"
1032	22/4	"		"	60	3500		"	"
1033	23/4	"		"	5	200		"	"
1034	"	"		"	60	4500		"	"
1035	24/4	"		"	55	4500		"	[illegible]

Luftkämpfe	Starts	Zahl der [illegible]	Feindflüge	Bemerkungen
1 mit 1 R.E. [illegible]			179	137. Jagdflug
			180	138. "
			181	139. "
			182	140. "
			183	141. "
		1		Überlandflug
		1		"
		1		"
		1		"
1 mit 1 B.H. Armentières			184	142. Jagdflug
			185	143. "

Below: Excerpt of the logbook of Lt. Hans Holthusen from the 15th April 1918 to 24th April 1918. On the 21th April 1918 he achived his first victory in an air combat against three S.E.5a fighters.

186 144. "

abgebr. "

ji 187 145. "

...ft 188 146. "

2 56 188

5134

Luftkämpfe	Stück	Zahl der Landungen		Feind-flüge	Bemerkungen
1 mit 1 RE [illegible]				179	137. Jagdflug
				180	138. "
				181	139. "
				182	140. "
				183	141. "
...[illegible]		1			Überlandflug
		1			"
		1			"
		1			"
...[illegible] 1 mit 1 BH Armentières				184	142. Jagdflug
				185	143. "
				186	144. "
					abgebr. "
ji				187	145. "
...ft				188	146. "
		2	56	188	

Below: Excerpt of the logbook of Lt. Hans Holthusen from the 6th July 1918 to the 13th July 1918. On the 10th July he achieved his second victory in an air combat against 17 Sopwith Camels.

Nr.	Datum		Führer	Beobachter	Flugweg	Dauer	Höhe	km	Maschine	Wetter
1111	6/7.	n.	selbst		6. Armee	15	500		Fokker D7 370	bedeckt
1112	"		"		"	85	5500		"	klar
1113	7/7.	n.	"		"	60	5400		"	"
1114	"		"		"	10	4000		"	"
1115	"	v.	"		"	50	4000		"	"
1116	8/7	n.	"		"	65	"		"	bedeckt
1117	"	v.	"		"	75	4300		"	klar
1118	"		"		"	60	4000		"	bedeckt
1119	9/7.	v.	"		"	45	3500		"	"
1120	10/7.	n.	"		"	50	4300		"	"
1121	11/7.	n.	"		"	55	4800		"	klar
1122	"	"	"		"	70	5000		"	[illegible]
1123	12/7.	v.	"		"	60	4500		"	[illegible]
1124	13/7.	n.	"		"	75	4800		"	[illegible] bedeckt
1125	"	n.	"		"	80	4000		"	"

Below: Excerpt of the logbook of Lt. Hans Holthusen from the 6th July 1918 to the 13th July 1918. On the 10th July he achieved his second victory in an air combat against 17 Sopwith Camels.

Luftkämpfe	Start	Zahl der Kämpfe	Zahl der Werkstattflüge	Frontflüge	Bemerkungen
					M.G. Schießen 125:60
1 mit 8 SE5 und Sopwith				239	197. Jagdflug
	ja			240	198. "
					M.G. Schießen 150:60
1 mit 5 Sopw. C. u. 2 B.F.				241	199. Jagdflug
1 mit 5 Sopwith 1 " 5 SE5 1 " 8 BF	ja			242	200. "
	"			243	201. "
1 mit 2 BF u. 5 Sopwith	"			244	202. "
1 mit 1 DH4	ja			245	203. "
1 mit 17 Sopwith 1 Abschuss (2.)	"			246	204. " [illegible] Camel ab
1 mit 3 SE5 [illegible]				247	205. "
				248	206. "
	ja			249	207. "
	"			250	208. "
1 mit 1 DH4 und 5 SE5	"			251	209. "
2		2			

Facing Page: *Desperate Maneuvers* by Russell Smith

This painting, commissioned specifically for this book, depicts the action of the last week of August 1917 described by Lt. Otto Fuchs to Bruno Schmäling on page 62.

Front Cover: *Jagdstaffel 30 Hunting Party* by Russell Smith

Lt. Otto Fuchs, Lt. Kurt Katzenstein and Oblt. Hans Bethge get ready to attack a formation of British aircraft over Vimy in September 1917.

Back Cover: *Jasta 30* by Jerry Boucher

Three Fokker D VII fighters of Jagdstaffel 30 in September 1918 flown by Lt. August Hartmann, Lt. Hans-Georg von der Marwitz and Lt. Hans Holthusen gaining height on their way to intercept British aircraft.

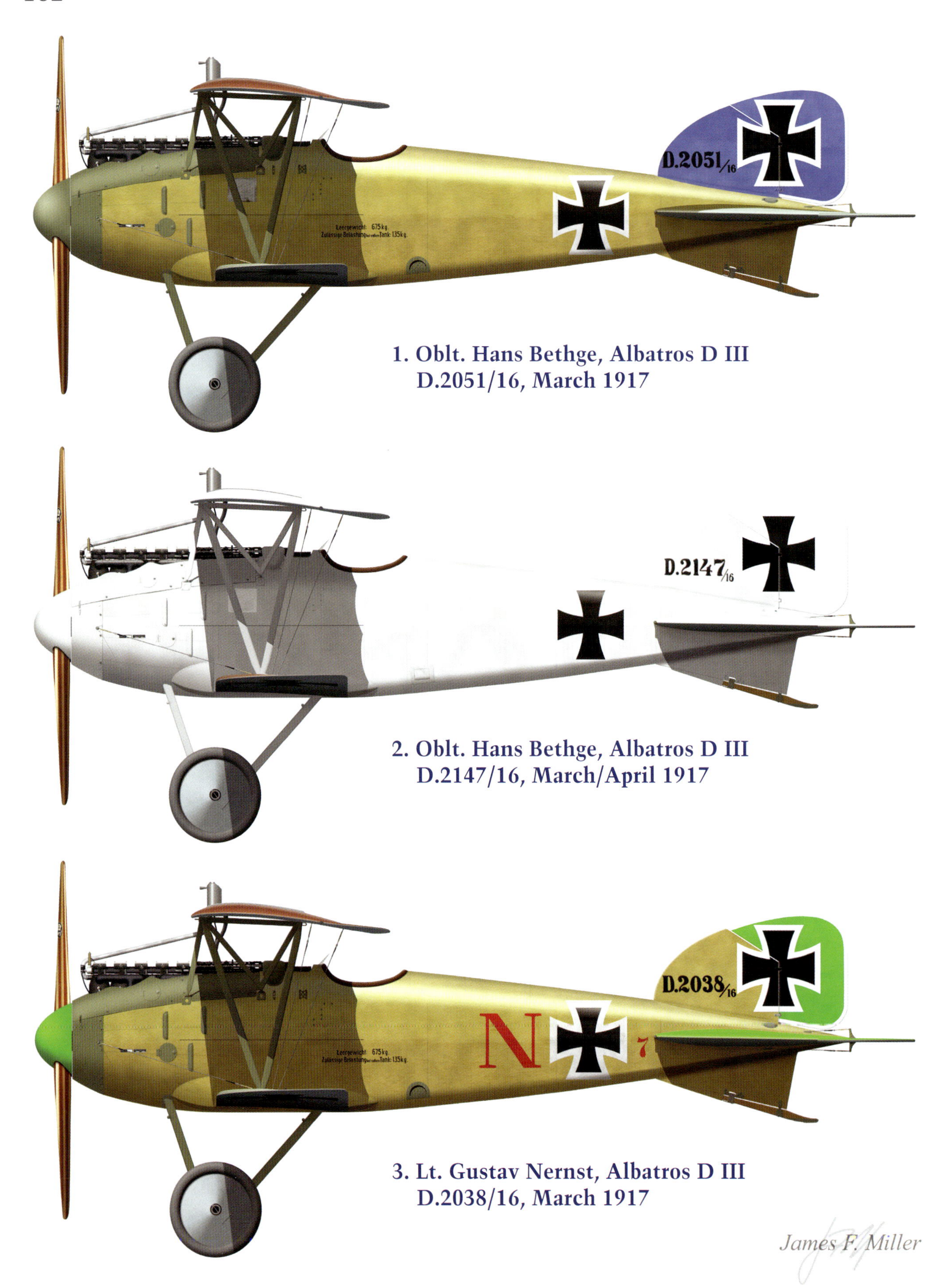

1. Oblt. Hans Bethge, Albatros D III D.2051/16, March 1917

2. Oblt. Hans Bethge, Albatros D III D.2147/16, March/April 1917

3. Lt. Gustav Nernst, Albatros D III D.2038/16, March 1917

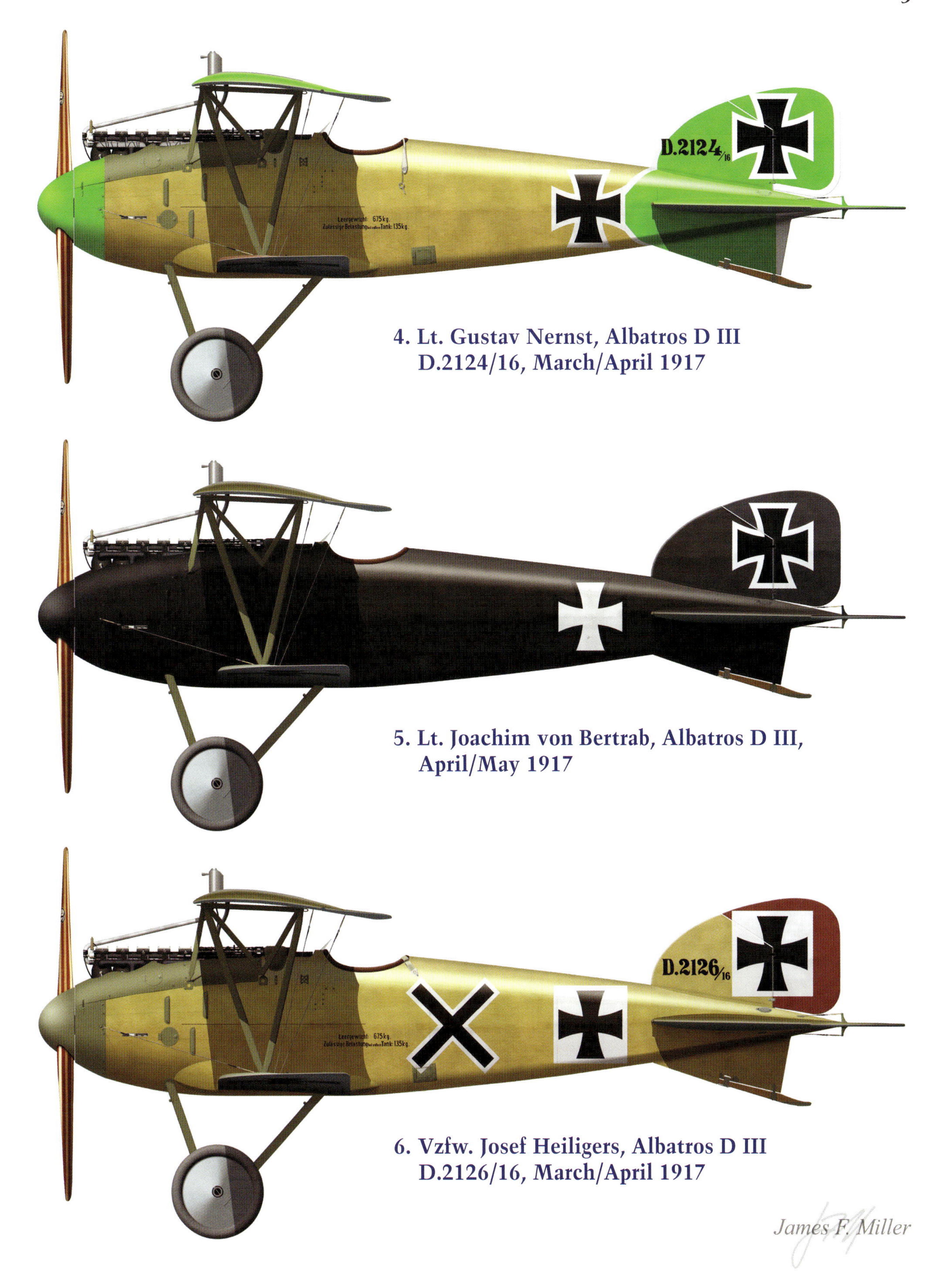

4. Lt. Gustav Nernst, Albatros D III D.2124/16, March/April 1917

5. Lt. Joachim von Bertrab, Albatros D III, April/May 1917

6. Vzfw. Josef Heiligers, Albatros D III D.2126/16, March/April 1917

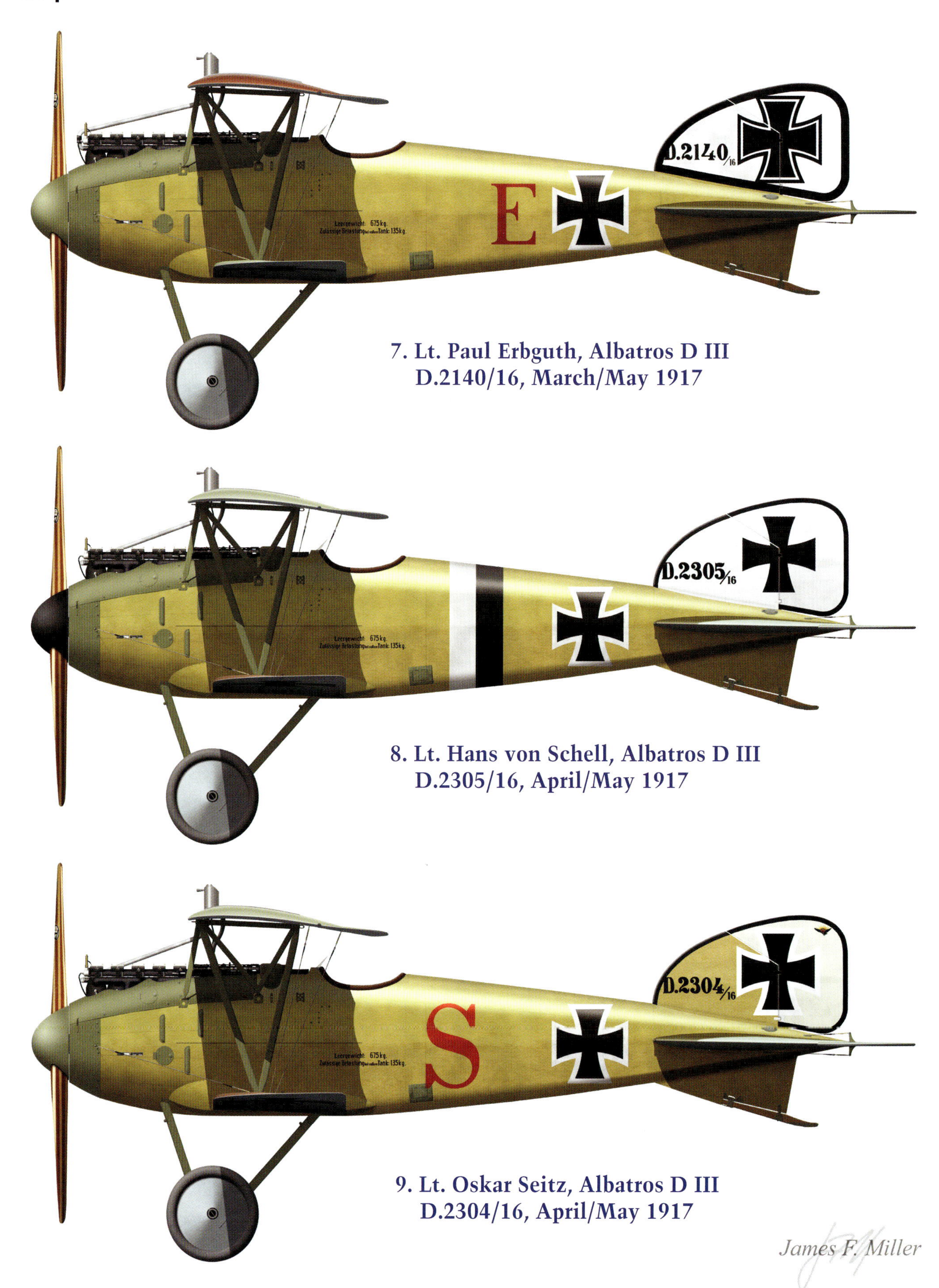

7. Lt. Paul Erbguth, Albatros D III
D.2140/16, March/May 1917

8. Lt. Hans von Schell, Albatros D III
D.2305/16, April/May 1917

9. Lt. Oskar Seitz, Albatros D III
D.2304/16, April/May 1917

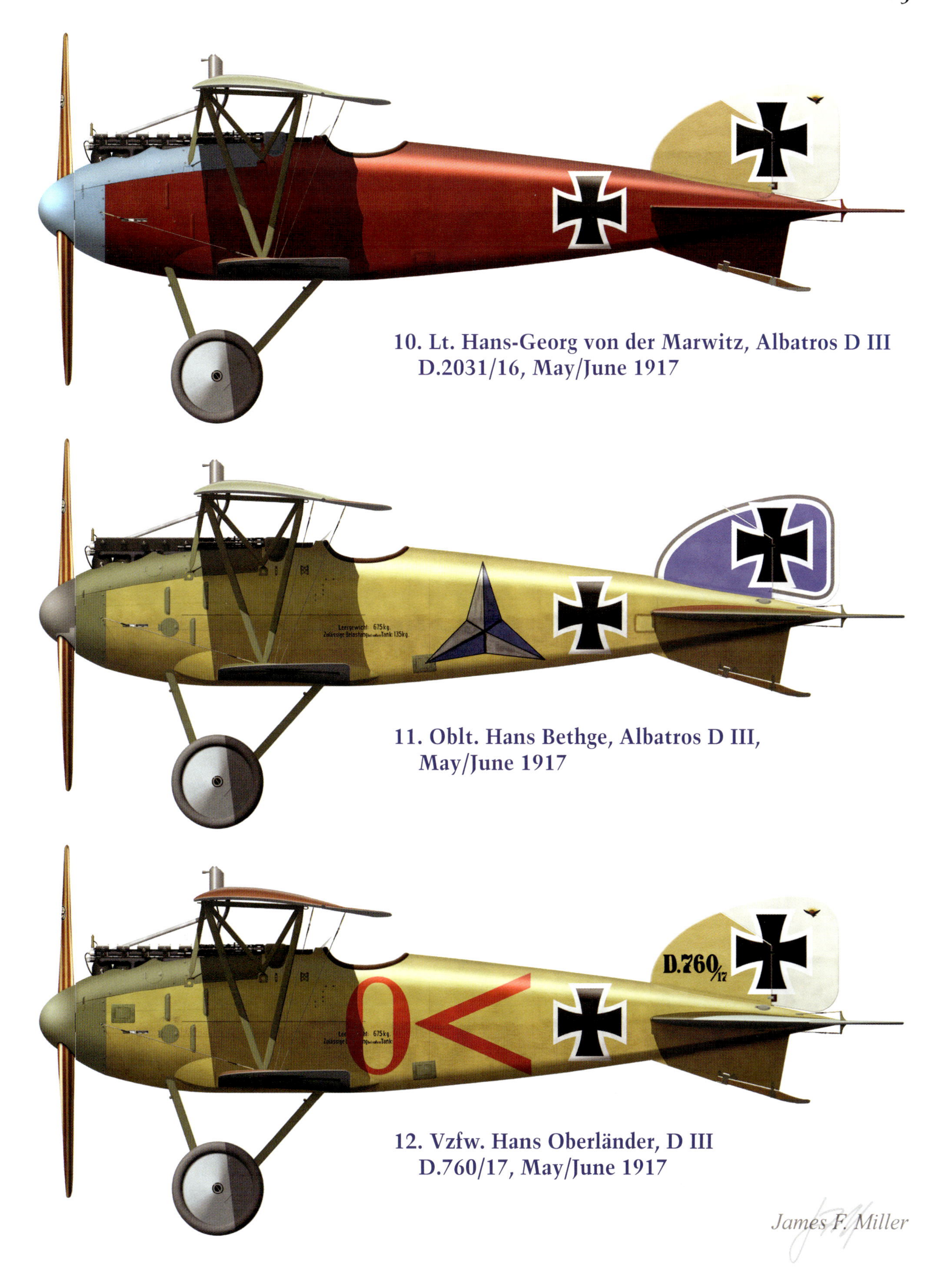

10. Lt. Hans-Georg von der Marwitz, Albatros D III D.2031/16, May/June 1917

11. Oblt. Hans Bethge, Albatros D III, May/June 1917

12. Vzfw. Hans Oberländer, D III D.760/17, May/June 1917

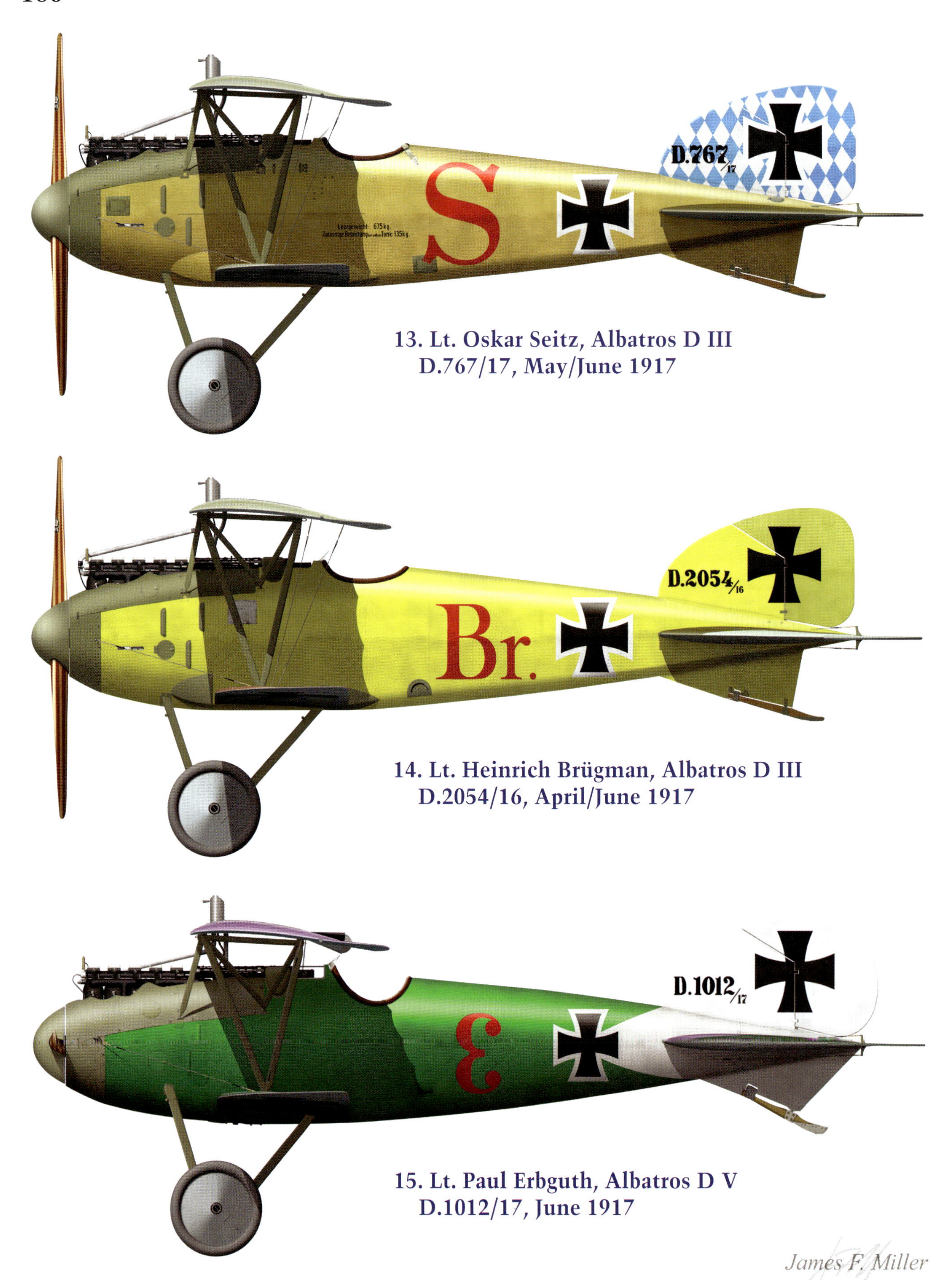

13. Lt. Oskar Seitz, Albatros D III D.767/17, May/June 1917

14. Lt. Heinrich Brügman, Albatros D III D.2054/16, April/June 1917

15. Lt. Paul Erbguth, Albatros D V D.1012/17, June 1917

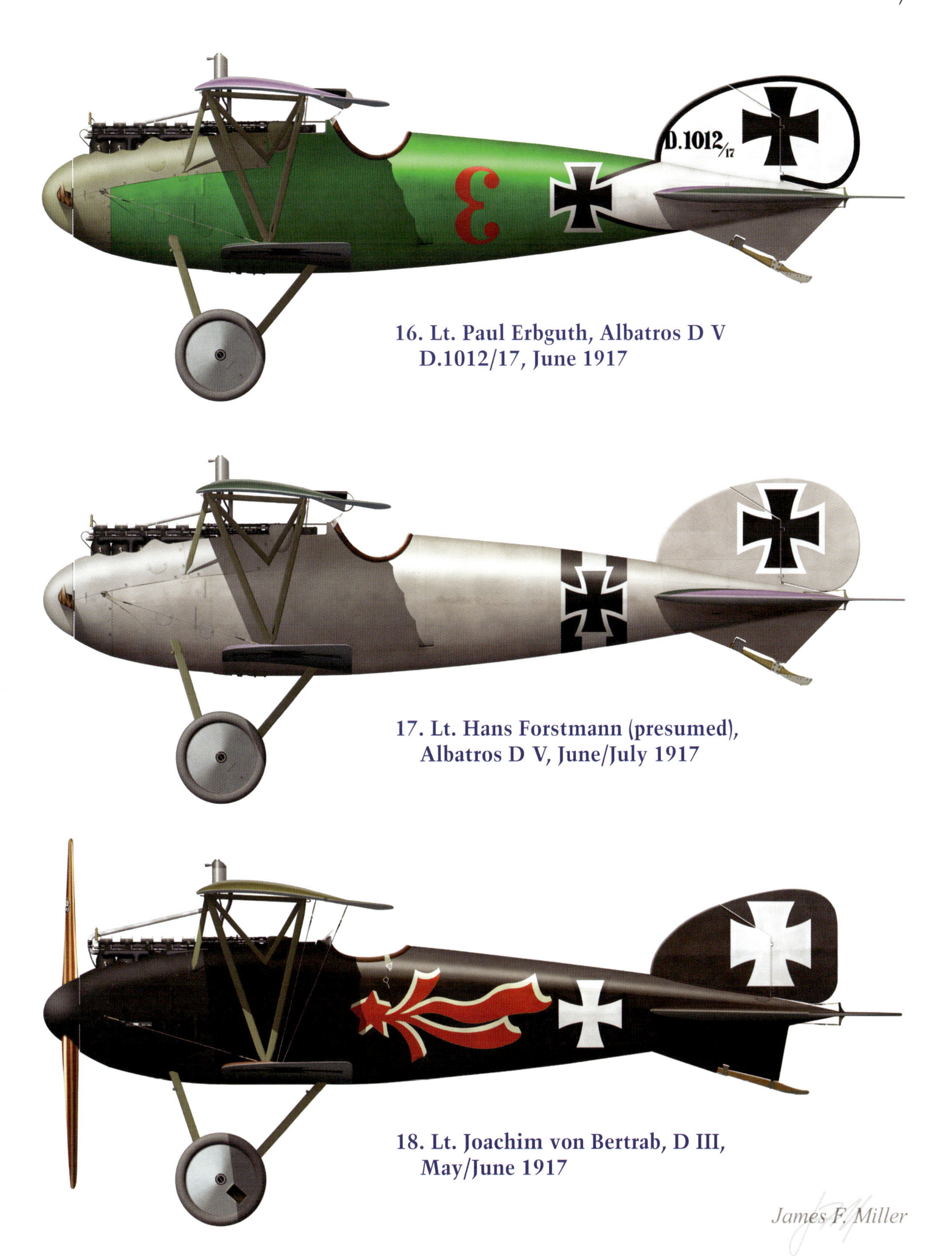

16. Lt. Paul Erbguth, Albatros D V D.1012/17, June 1917

17. Lt. Hans Forstmann (presumed), Albatros D V, June/July 1917

18. Lt. Joachim von Bertrab, D III, May/June 1917

19. Lt. Joachim von Bertrab, Albatros D V, July/August 1917

20. Lt. Hans-Georg von der Marwitz, Albatros D V D.1016/17, June 1917

21. Lt. Otto Fuchs, Albatros D III, June 1917

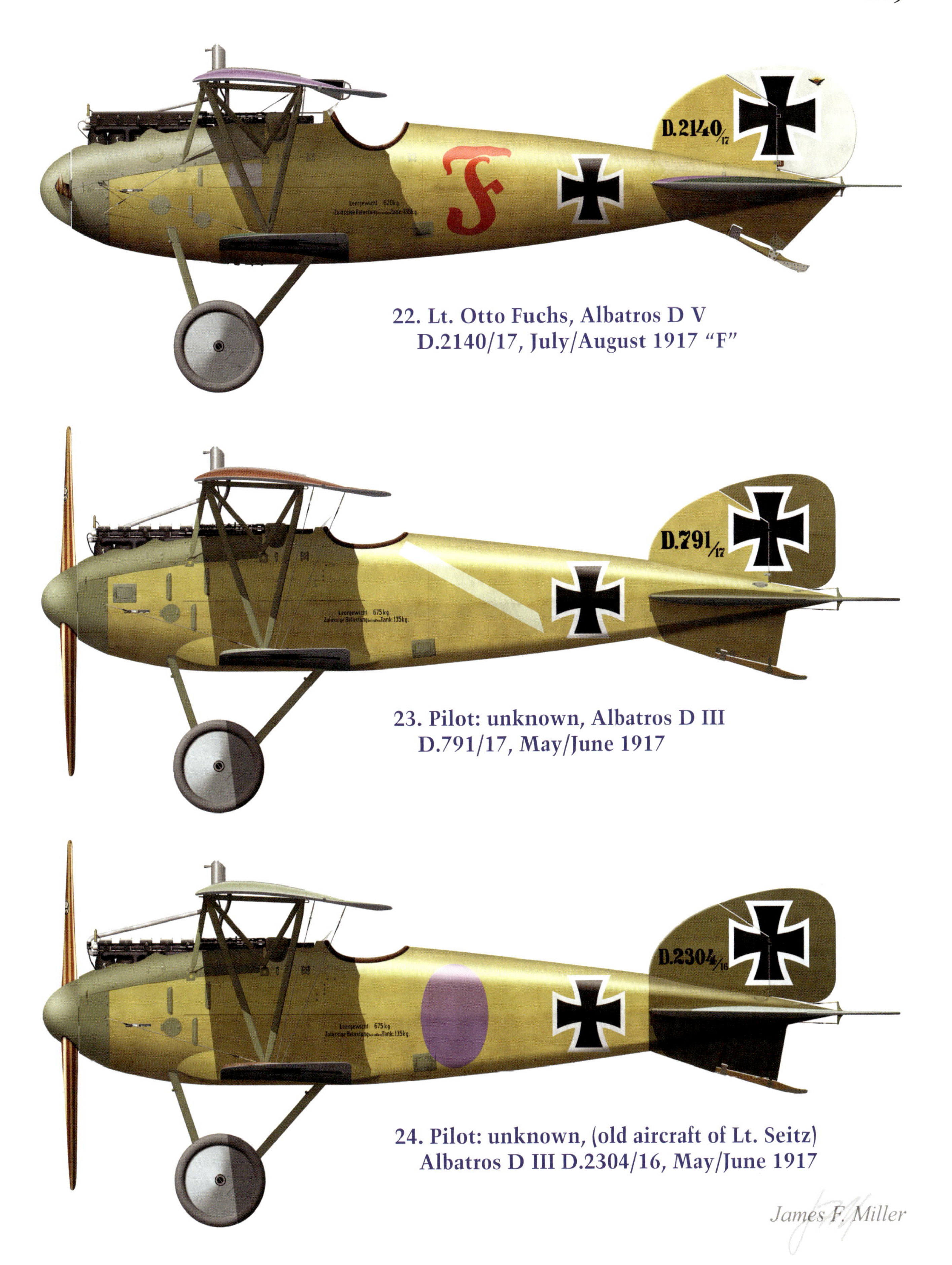

22. Lt. Otto Fuchs, Albatros D V D.2140/17, July/August 1917 "F"

23. Pilot: unknown, Albatros D III D.791/17, May/June 1917

24. Pilot: unknown, (old aircraft of Lt. Seitz) Albatros D III D.2304/16, May/June 1917

James F. Miller

25. Lt. Otto Fuchs, Albatros D V, September/October 1917 (Reconstruction according to information from Otto Fuchs)

26. Lt. Hans-Georg von der Marwitz, Albatros D V, August/October 1917

27. Lt. Kurt Katzenstein, Albatros D V, August/October 1917

James F. Miller

28. Lt. Rudolf von der Horst, Albatros D V, August/October 1917

29. Lt. Hans Holthusen, Albatros D V, October 1917 (Reconstruction according to information from Hans Holthusen)

30. Lt. Karl Weltz, Albatros D V, September/October 1917 (Reconstruction according to information from Otto Fuchs)

James F. Miller

31. Vzfw. Josef Heiligers, Albatros D V, July/September 1917

32. Oblt. Hans Bethge, Albatros D V, July/October 1917

33. Lt. Hans-Georg von der Marwitz, Albatros D V, November 1917/February 1918

34. Lt. Karl Weltz, Albatros D V D.4420/17, September/October 1917

35. Uffz. Emil Liebert, Albatros D V, November 1917/January 1918

36. Vzfw. Josef Heiligers, Albatros D V, November/December 1917

37. Lt. Hans Holthusen, Albatros D V,
November 1917/February 1918
(Reconstruction according to
information from Hans Holthusen)

38. Lt. Wendel Bastgen, Albatros D V,
January/February 1918
(Reconstruction according
to the British G-Report)

39. Oblt. Hans Bethge, Pfalz D IIIa
4203/17, February/March 1918

James F. Miller

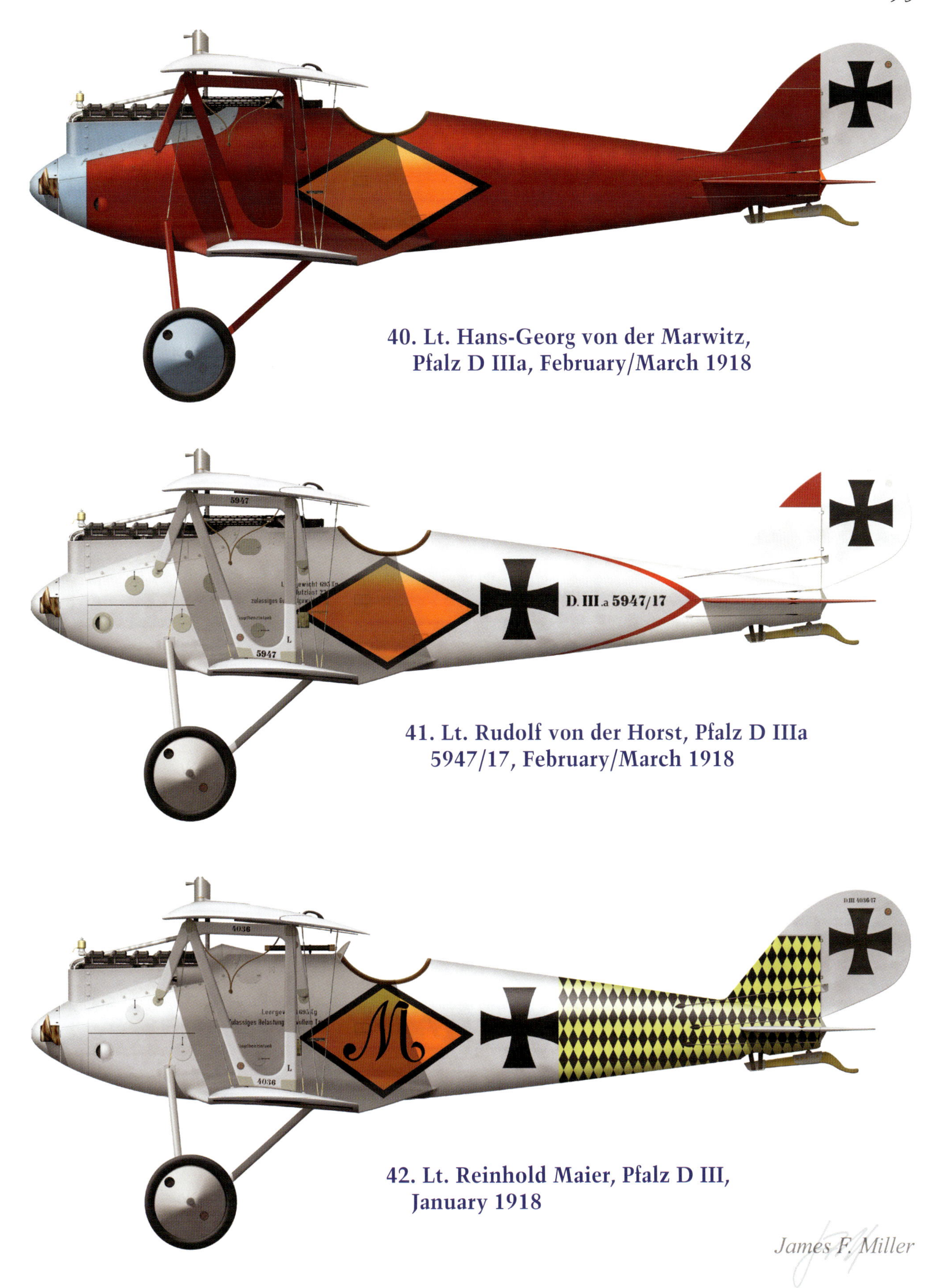

40. Lt. Hans-Georg von der Marwitz, Pfalz D IIIa, February/March 1918

41. Lt. Rudolf von der Horst, Pfalz D IIIa 5947/17, February/March 1918

42. Lt. Reinhold Maier, Pfalz D III, January 1918

James F. Miller

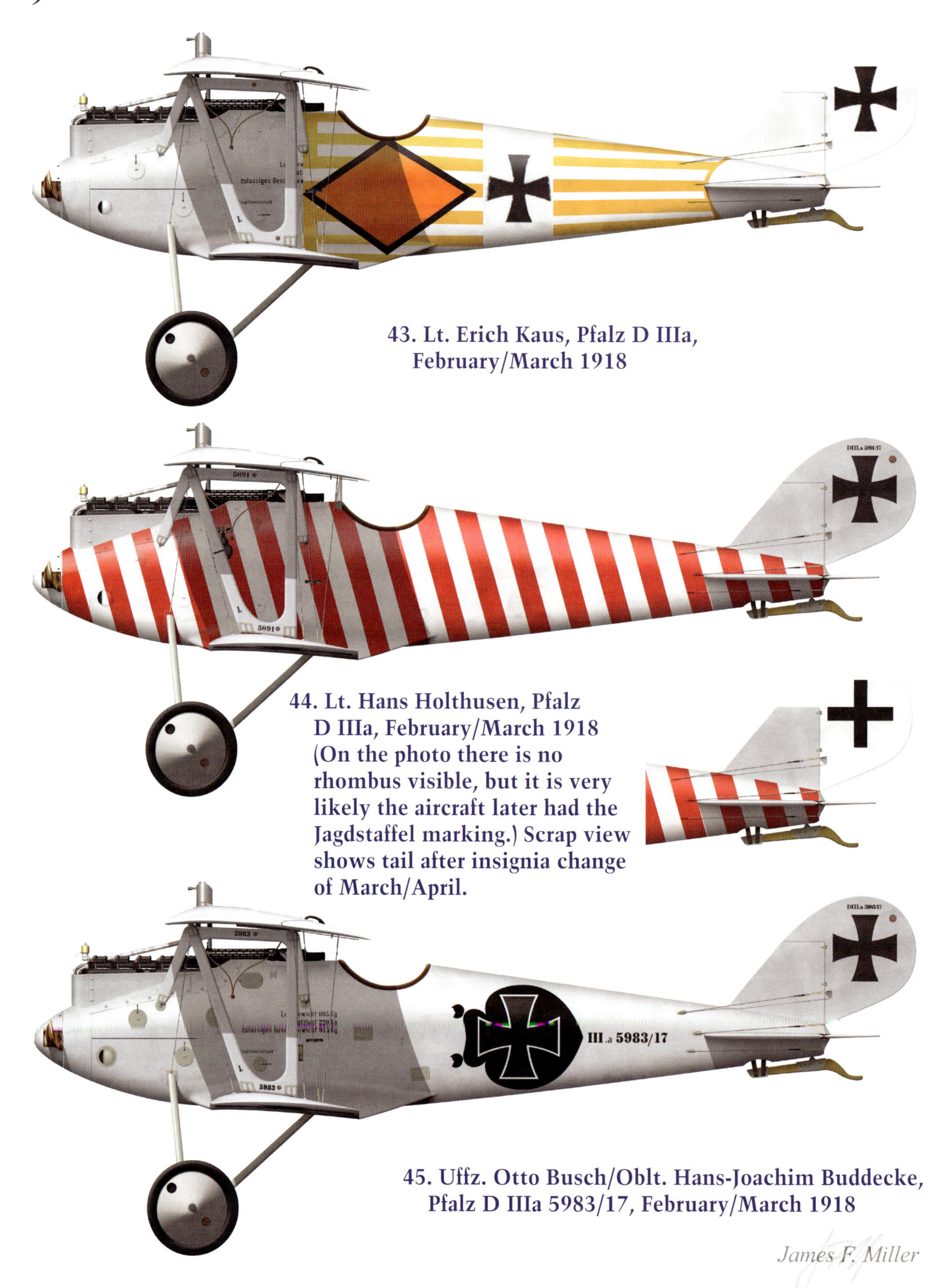

43. Lt. Erich Kaus, Pfalz D IIIa, February/March 1918

44. Lt. Hans Holthusen, Pfalz D IIIa, February/March 1918 (On the photo there is no rhombus visible, but it is very likely the aircraft later had the Jagdstaffel marking.) Scrap view shows tail after insignia change of March/April.

45. Uffz. Otto Busch/Oblt. Hans-Joachim Buddecke, Pfalz D IIIa 5983/17, February/March 1918

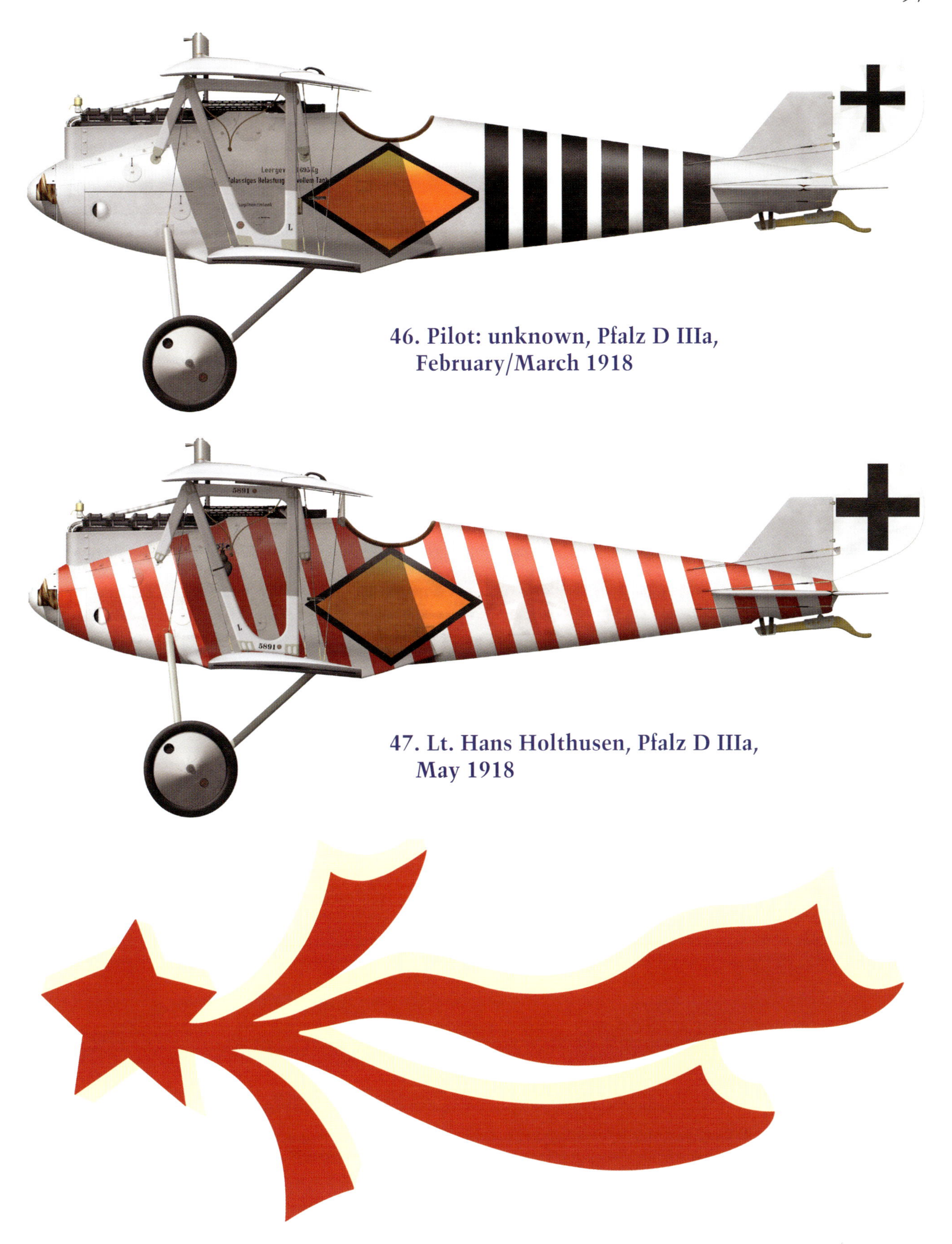

46. Pilot: unknown, Pfalz D IIIa, February/March 1918

47. Lt. Hans Holthusen, Pfalz D IIIa, May 1918

Detail of Bertrab's Albatros D III Artillery Shell

James F. Miller

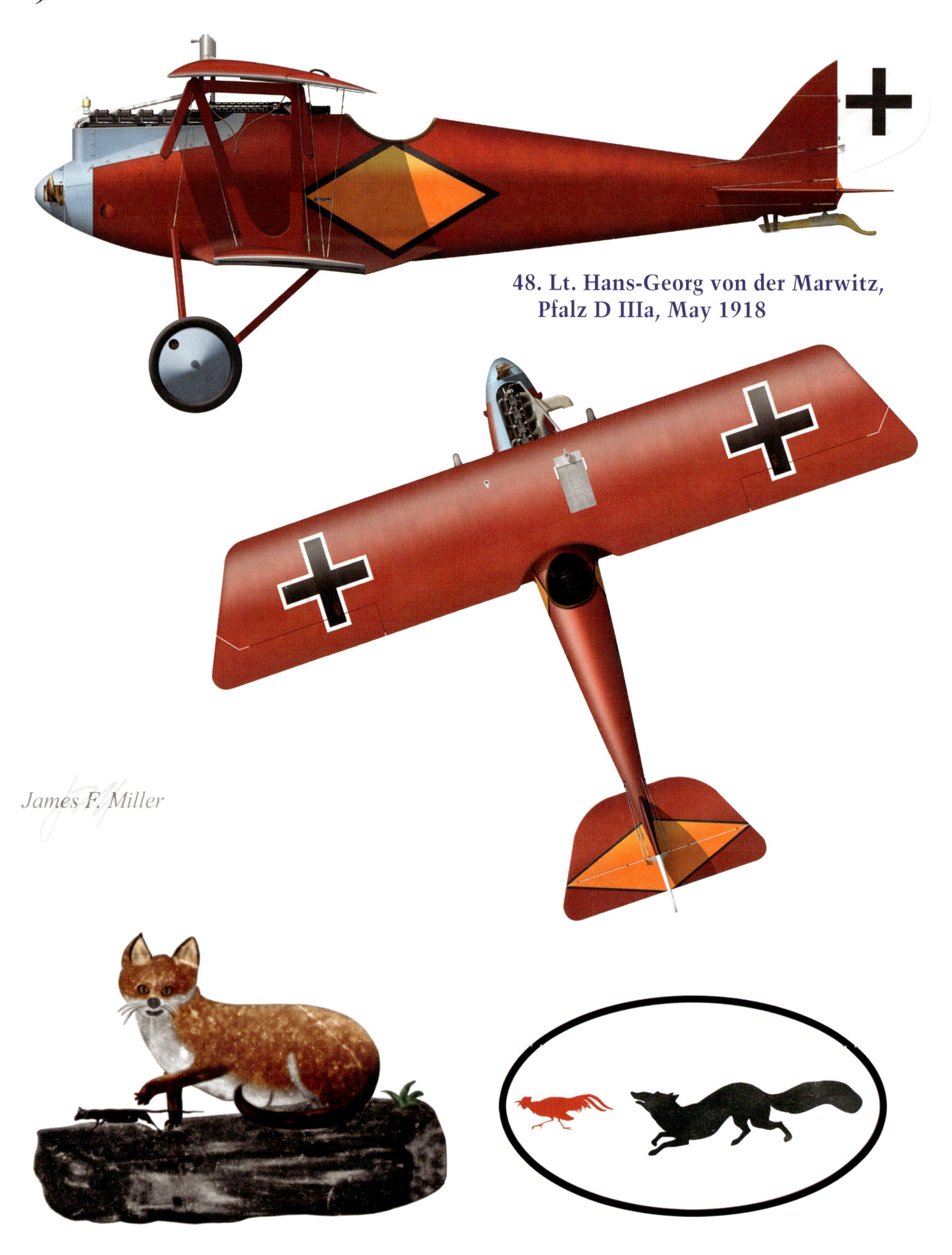

48. Lt. Hans-Georg von der Marwitz, Pfalz D IIIa, May 1918

Katzenstein Albatros D V Detail

Fuchs Verdigris Green Albatros D V Detail

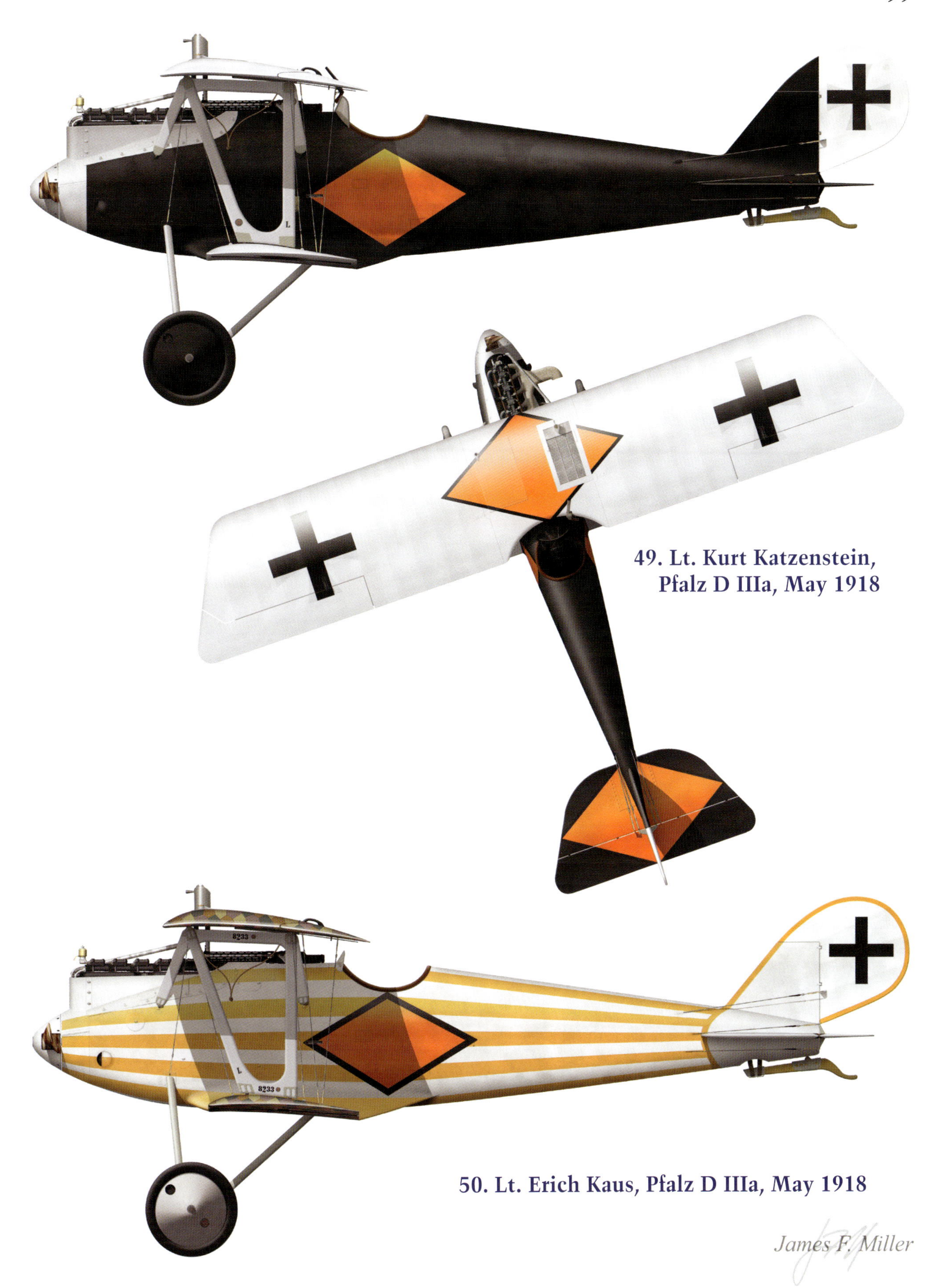

49. Lt. Kurt Katzenstein, Pfalz D IIIa, May 1918

50. Lt. Erich Kaus, Pfalz D IIIa, May 1918

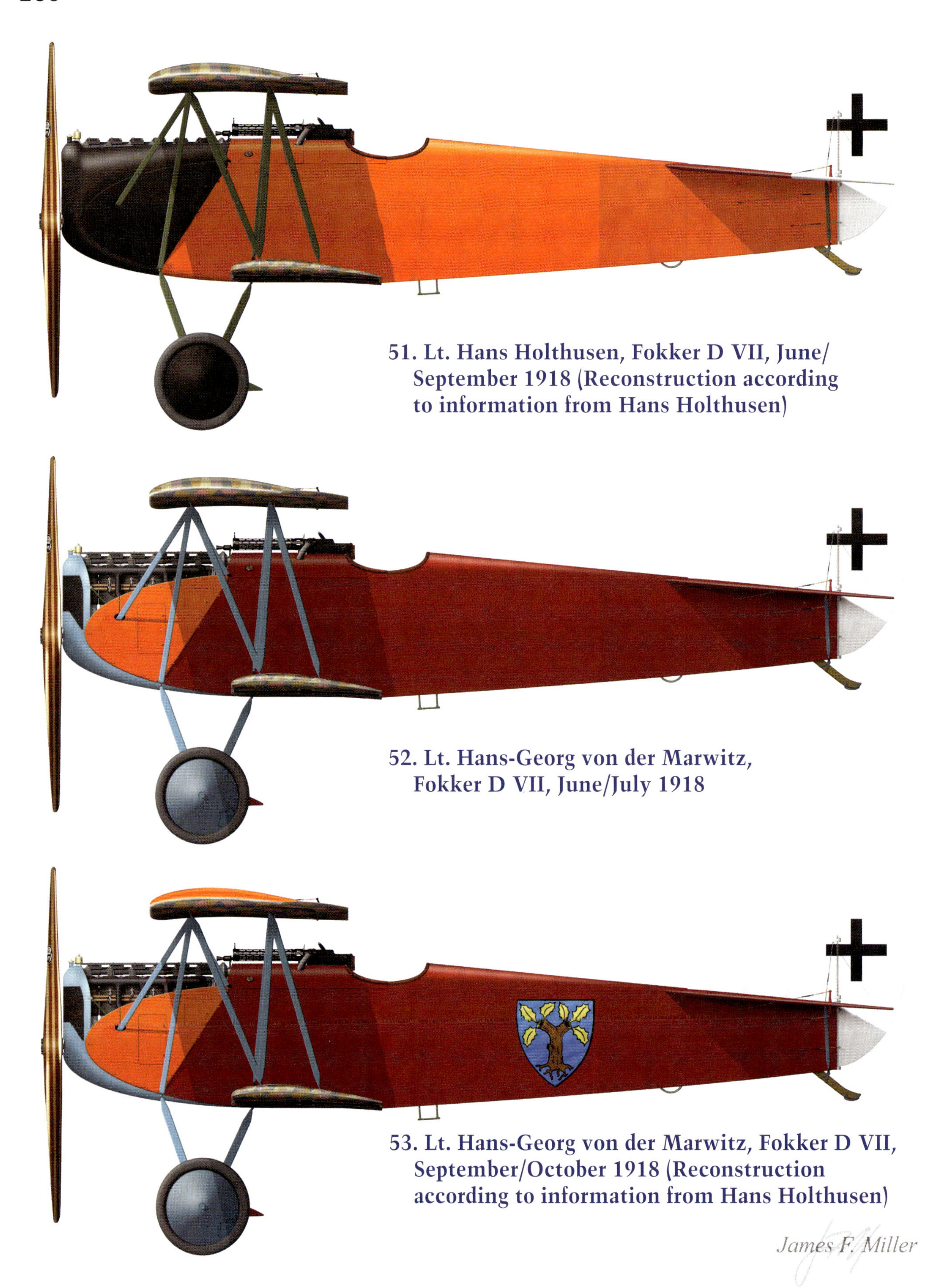

51. Lt. Hans Holthusen, Fokker D VII, June/September 1918 (Reconstruction according to information from Hans Holthusen)

52. Lt. Hans-Georg von der Marwitz, Fokker D VII, June/July 1918

53. Lt. Hans-Georg von der Marwitz, Fokker D VII, September/October 1918 (Reconstruction according to information from Hans Holthusen)

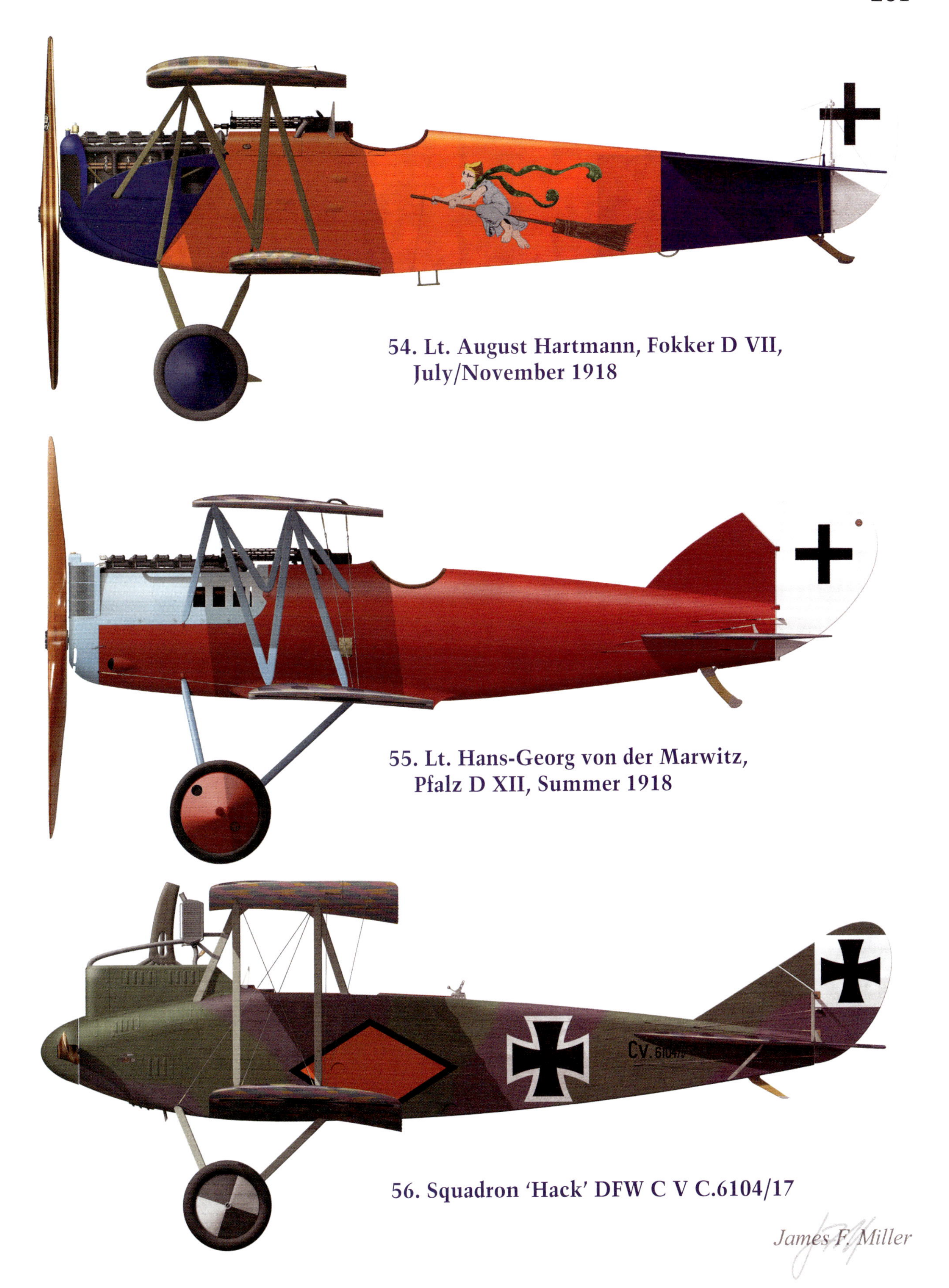

54. Lt. August Hartmann, Fokker D VII, July/November 1918

55. Lt. Hans-Georg von der Marwitz, Pfalz D XII, Summer 1918

56. Squadron 'Hack' DFW C V C.6104/17

Regimental Colors of Some Jasta 30 Pilots

Feld. Art. Reg. 46, von Bertrab

Ulanan Reg. 16,
von der Marwitz

Kürassier Reg. 8, Lt. Nernst

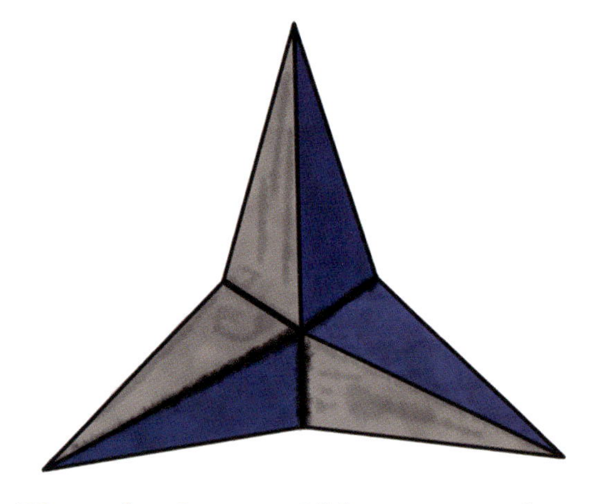

Mercedes Star on Oblt. Hans Bethge's
Albatros D III, May/June 1917

Brugman Albatros D III D.2054/16 Detail

James F. Miller

Fuchs Albatros D V D.2140/17 Detail

Marwitz Albatros D V D.1016/17 Detail

Heiligers Albatros
D III D.2126/16
Detail

Maier Pfalz
D III
4036/17
'M' Detail

Above & Below: Detail of "moi Hexle" (my little witch) on Lt. August Hartmann's Fokker D VII during July/November 1918. Moi Hexle was Hartmann's name for his aircraft.

Oberländer Albatros D III D.760/17 Detail

Marwitz Fokker D VII
Wappen Detail

Seitz Albatros D III D.767/17 S Detail

Seitz Albatros D III D.2304/16 Detail

Erbguth Albatros D III D.2140/16 Detail

James F. Miller

N 

Nernst Albatros D III D.2038/16 Detail

Erbguth Albatros D V D.1012/17 Detail

Bibliography

Published Sources

Beedle, J. *43 Squadron: The History of the Fighting Cocks, 1916–1966*. London: Beaumont, 1966.

Fuchs, Otto. *Wir Flieger: Kriegserinnerungen eines Unbekannten*. Leipzig: Koehler, 1933.

Henshaw, Trevor. *The Sky Their Battlefield*. London: Grub Street, 1995.

Imrie, Alex. *German Fighter Units 1914–May 1917*. London: Osprey, 1978.

Imrie, Alex. *German Fighter Units June 1917–1918*. London: Osprey, 1978.

Karten des französisch-belgischen Kriegsschauplatzes, 8th ed. Bielefeld: Velhagen & Klasing, 1915

Mars Wona Taschenatlas vom Westlichen Kriegsschaufplatz, Königswartha, Sachsen

Puglisi, Bill, ed. "Josef Raesch of Jasta 43." *The Cross & Cockade Journal* 8.4 (1967).

Richter, Georg. *Der königlich sächsische Militär St. Heinrichs Orden 1736–1918*. Frankfurt: Weidlich, 1964.

Rogers, Les. "RFC & RAF Casualties, 1917–1918." *Cross & Cockade (Great Britain)*, various issues.

Rogers, Les. "RNAS Casualties, 1916–1918." *Cross & Cockade (Great Britain)*, 15.4 (1984).

Stegemann, Hermann. *Geschichte des Krieges*, Band IV. Stuttgart: Deutsche Verlags-Anstalt, 1921.

Unpublished Sources

Archival Records

Daily flight reports, Intelligence and Tracking Station of A.O.K. 6, Bundles 13, 305, 323, Bavarian Main State Archive, Munich

Flight inventory list of Armee-Flug-Park 6, Bavarian Main State Archive, Munich

Flight Log Lt. Kurt Katzenstein, Stiftung Technikmuseum Berlin

Photo album Lt. Kurt Katzenstein, Stiftung Technikmuseum Berlin.

Personnel inventory of Jagdstaffel 30, June 1917 and October 1917, Bavarian Main State Archive, Munich

Personnel records of Otto Fuchs, OP 24636, Bavarian Main State Archive, Munich

Personnel records of Oskar Seitz, OP 12233, Bavarian Main State Archive, Munich

Rank lists and personnel rosters of Armee-Flug-Park 6, Bavarian Main State Archive, Munich

Rank lists and personnel rosters of Jagdstaffel 30 pilots: Bavarian Main State Archive, Munich; State Archive of Baden-Württemberg, Stuttgart; General State Archive, Karlsruhe

Weekly reports of the Kommandeur der Flieger of the 6th Army, Bavarian Main State Archive, Munich

Winfried Bock Collection

Excerpt from the War Diary of Jagdstaffel 30 (Transcript by Erich Tornuss with additions and corrections by Dr. Gustav Bock and Winfried Bock)

Nachrichtenblätter der Luftstreitkräfte

Handwritten notes by Dr. Gustav Bock from conversations with Hans Oberländer

Correspondence of Dr. Gustav Bock with Erich Kaus

Documents of Hans Holthusen

Combat Reports of 40 and 1 Squadrons, RFC/RAF

Draft writings re individual English squadrons by Paul Chamberlain or P.S. Brown

Alex Imrie Collection

Photos and documents of August Hartmann

Handwritten notes from conversations with Erich Kaus

Reconstruction of aircraft painting schemes (sketches) based on conversations with Erich Kaus

G-Report of captured German aircraft, February 16, 1918

Bruno Schmäling Collection

Otto Fuchs, *Wir Flieger*, unpublished partial manuscript

Otto Fuchs corrections and additions to book and manuscript *Wir Flieger*

Handwritten notes form the conversation with Otto Fuchs

Correspondence with Otto Fuchs

Photos of Otto Fuchs

Reconstruction of aircraft painting schemes (sketches) based on conversations with Otto Fuchs

Flight logbook and documents of Hans Holthusen

Handwritten notes from conversations with Hans Holthusen

Correspondence with Hans Holthusen

Reconstruction of aircraft painting schemes (sketches) based on conversations with Hans Holthusen

Photos and documents of Erich Kaus

Handwritten notes from conversation with Erich Kaus

Photos and documents of Rudolf Freiherr von der Horst zu Hollwinkel

Photos and documents of Paul Erbguth, Erbguth family

Documents of Hans-Joachim von Bertrab by Herr Olaf von Bertrab

Diary of Josef Raesch, 1917–1918 (original copy)

Reinhard Zankl Collection

Documents regarding the aircraft inventory of Jagdstaffel 30

Index

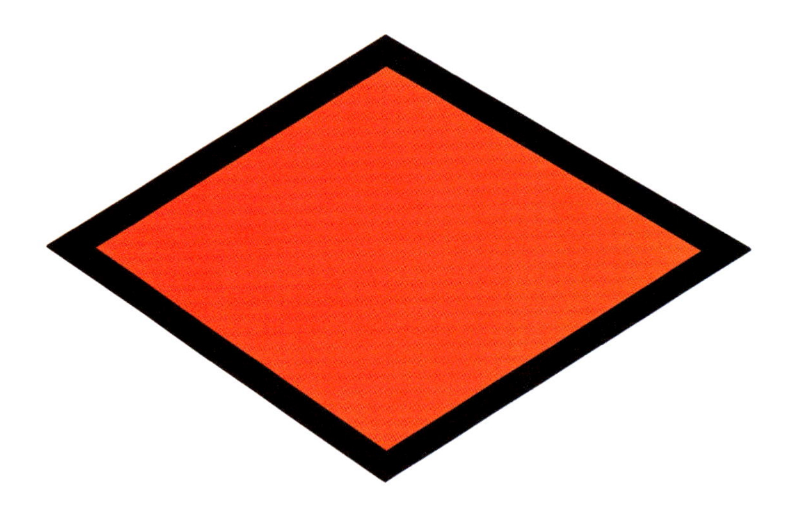

Jasta 30 Diamond

Uffz. Otto Busch/Oblt. Hans-Joachim Buddecke, Pfalz D IIIa 5983/17, Personal Insignia